Biological control of insects and mites

An introduction
to beneficial natural enemies
and their use in pest management

preface

Why use biological control?

Since the late 1940s, insect control has relied heavily on synthetic chemical insecticides. Insecticides are relatively easy to use and have generally provided safe and effective pest control. They will certainly continue to be a component of most pest management programs. Many newer pesticides made available in the past decade or so are more selective and less hazardous than most of the older compounds. Nevertheless, most insecticides have at least some of these undesirable attributes: they usually present some degree of hazard to the applicator and other people who may come in contact with them; they can leave residues that some find unacceptable; they can contaminate soil and water and affect wildlife, aquatic life, and other nontarget organisms; they can interfere with beneficial organisms, such as pollinating insects and the natural enemies of pests; and insects can develop resistance to insecticides, effectively eliminating those materials as pest management options. In addition, organic standards prevent the small but rapidly growing number of organic growers and processors from using synthetic chemicals. For these reasons, many farmers and gardeners are exploring and adopting methods that reduce pesticide use.

Biological control represents one alternative to the use of insecticides. Biological control is the conscious use of living beneficial organisms, called natural enemies, for the control of pests. Virtually all pests have natural enemies and appropriate management of natural enemies can effectively control many pests. Although biological control will not control all pests all of the time, it should be the foundation of an approach called integrated pest management, which combines a variety of pest control methods. Biological control can be effective, economical, and safe, and it should be more widely used than it is today.

Why this publication?

Biological control relies on living organisms that must have food and shelter and that interact with the pests, the crop, and other environmental factors. The pest manager (a farmer, crop consultant, or gardener) should be able to recognize important natural enemies, understand their needs, and know how to maximize their effectiveness. This requires different knowledge and skills than are needed for chemical control. A good understanding of the relationships between pests, their natural enemies, and the environment is essential for success in biological control. The need for this type of knowledge is the rationale for this publication.

This publication provides an introduction to the biological control of pest insects and mites. Because successful biological control relies on knowledge of pests and their natural enemies, we include basic biological information on insects and discuss how insects become pests. We also discuss biological control in the context of other forms of pest control, examining the role of the environment in suppressing pests (natural control), as well as the various general methods for controlling insects.

A major portion of the publication is devoted to a survey of natural enemies. There are hundreds of important natural enemies in the North Central United States, and it is impractical to cover all of them in this publication. Instead, we provide examples from the more common groups of insect predators, parasitic insects, insect-parasitic nematodes, and insect pathogens. We include numerous photographs to help you recognize these beneficial organisms and suggest several sources of further information and images.

The remainder of the publication discusses the three broad approaches to biological control of insects: finding new useful natural enemies, enhancing the effectiveness of natural enemies by protecting them from harm and providing them with necessary habitat and other resources, and releasing additional natural enemies when those naturally present are not adequate.

Numerous books and thousands of scientific and nontechnical articles have been written on this subject. This publication is not an encyclopedic summary of all this information. Instead, we hope this overview stimulates you to seek additional information on the use of natural enemies for your specific pest control needs. Additional Extension resources relevant to the North Central states include biological control publications on pests of cabbage and related crops (NCR471) and greenhouse crops (NCR531) produced by the University of Wisconsin, and pests of field crops (E-2721), forests (E-2679), and home gardens (E-2719) produced by Michigan State University.

The first edition of this publication, now out of print, was printed in 1993. In the intervening years there have been many new developments in biological control, including the arrival of new pests such as soybean aphid, the development of new classes of insecticides that are less harmful to beneficial natural enemies, and the production of new resource materials available to farmers, foresters, and gardeners. The original authors (Mahr and Ridgway) thank Michelle Miller, Center for Integrated Agricultural Systems, College of Agricultural and Life Sciences, University of Wisconsin–Madison, for her dedication to the production of this revised edition, and Dr. Paul Whitaker, Department of Biological Sciences, University of Wisconsin–Marathon County, for researching and writing the new material.

I hope this publication helps you plan and conduct successful biological control.

Daniel L. Mahr, Coordinator
  Extension Biological Control Programs
  Department of Entomology
  University of Wisconsin-Madison
  January 2008

acknowledgements

This publication was funded, in part, by University of Wisconsin-Extension, the Wisconsin Institute of Sustainable Agriculture of the University of Wisconsin–Madison, the University of Wisconsin–Marathon County, and the North Central Region Sustainable Agriculture Research and Education (SARE) program of the United States Department of Agriculture, Cooperative Extension Service. The first edition of this publication was substantially funded by a grant from the Pesticide Impact Assessment Program of the United States Department of Agriculture's Cooperative Extension Service (now the Cooperative State Research, Education, and Extension Service).

We thank 11 reviewers for their insightful comments and suggestions on this second edition. They are farmers Dan Specht, Harry Hoch, Linda Grice, Linda Halley, Kevin Shelley, and Sandra Whitney; and researchers Doug Landis, Michigan State University; Mike Hogan, Ohio State University–Extension; Rick Weinzierl, University of Illinois at Champaign-Urbana; Bob Wright, University of Nebraska–Lincoln; and Cliff Sadof, Purdue University. We also thank Mimi Broeske, Nutrient and Pest Management Program, University of Wisconsin–Madison, for suggestions on content. And we thank Tom Kalb, Kenosha County Cooperative Extension Office, Kenosha, Wisconsin; Susan Mahr, University of Wisconsin–Madison; and Rick Weinzierl and Bob Wright for reviewing and providing comments on the first version of the manuscript. The anonymous reviews of North Central States Extension entomologists were also appreciated.

This publication would be much less useful without the many excellent photographs of natural enemies that were supplied by several people. The contributors for each photograph are acknowledged at the end of the publica-tion. We also thank the several people who supplied excellent photographs that we were unable to use. Additionally, we would like to thank Kerry Katovich, Susan Mahr, Steve Mroczkiewicz, Ken Raffa, Kammy Schell, Mike Strand, Dave Hogg, and Merritt Singleton, all of the Department of Entomology, University of Wisconsin–Madison, for providing living insects for photographic purposes. We appreciate the loan of museum specimens by Steven Krauth from the Insect Research Collection, Department of Entomology, University of Wisconsin–Madison, for illustration purposes. Commercial natural enemies were kindly provided for photographic purposes by Applied Bionomics Ltd., Sidney, BC; Gardens Alive!, Lawrenceburg, IN; IPM Laboratories, Inc., Locke, NY; and Biotactics, Romoland, CA.

We are grateful to Diana Budde of the art department at the University of Wisconsin–Marathon County and art student Abigail Guzinski for creating all of the line drawings in this publication on relatively short notice.

We acknowledge with gratitude the support, encouragement, and patience of colleagues within the Department of Entomology and the Wisconsin Institute of Sustainable Agriculture at the University of Wisconsin–Madison, and members of the North Central Regional Committee on the Biological Control of Pest Arthropods (NCR125), for their work to bring the first and second editions to fruition.

The authors of the first edition acknowledge the encouragement and support of Michelle Miller for this revised edition, and the authorship of Paul Whitaker for the expanded and updated text.

The creativity, hard work, and good humor of the Cooperative Extension Publications Unit of the University of Wisconsin-Extension, especially Linda Deith and Susan Anderson, is deeply appreciated.

A NOTE ABOUT SCIENTIFIC NAMES
Most natural enemies do not have common names. Therefore, we use scientific names throughout the publication. The section entitled "Classification" on page 8 provides an explanation of the nature and use of scientific names.

TERMS IN BOLDFACE
Important terms are printed in **boldface** where they first appear in the text and are defined in the Glossary (pages 103–105).

contents

What is biological control?

Overview

Biological control is the intentional manipulation of populations of living beneficial organisms in order to limit populations of pests. Although biological control can be used to control weeds, the microorganisms that cause plant diseases, and even some vertebrates, this publication focuses on the biological control of insects and mites. In this publication, we refer to the beneficial organisms that attack pests as **natural enemies**, though you may also know them as "good bugs," beneficial insects, or beneficials. The natural enemies of insects are a diverse group of organisms that includes predators, parasitic insects, nematodes, and various microorganisms. The intent of biological control is not to eradicate pests, but to keep them at tolerable levels at which they cause no appreciable harm. In fact, because natural enemies require their prey or hosts for survival, biological control works best when there is always a small population of pests to sustain their natural enemies. This is a major difference between biological control and the use of pesticides.

There are three broad approaches to biological control:

- **Classical biological control** (also called **importation of natural enemies**) involves the importation, screening, and release of natural enemies to permanently establish effective natural enemies in new areas. Classical biological control usually targets introduced (**non-native**) pests, most of which arrive here without the natural enemies that control their populations in their native lands. Native pests that are not adequately controlled by existing natural enemies may also be the target of classical biological control. These activities are tightly regulated and are conducted solely by federal and state agencies, unlike the following two approaches, which can be used by anyone.

- **Augmentative biological control** (or augmentation of natural enemies) typically involves the purchase and release of natural enemies that are already present in the United States but may not be numerous enough to adequately control pests in a particular location. The goal of augmentative biological control is to temporarily increase the number of natural enemies and, therefore, the level of biological control of the target pest.

- **Conservation biological control** (or conservation of natural enemies) improves the effectiveness of natural enemies through farming and gardening practices that provide necessary resources for their survival and protect them from toxins and other adverse conditions. These conservation practices will benefit all natural enemies, whether they are native, successfully established through classical biological control, or released for augmentative biological control.

These three approaches are discussed in greater detail in chapters 7–9, but a few historical examples may help clarify the three approaches to using natural enemies in biological control.

Many centuries ago, Chinese farmers observed that ants were helping to control insect pests in their citrus orchards by feeding on caterpillars, beetles, and leaf-feeding bugs. The farmers discovered that collecting the papery nests of these ants from trees in the countryside and moving them into their orchards improved control of some orchard pests. They also provided aerial bamboo runways among the citrus trees to help the ants move easily from tree to tree. These efforts to increase the numbers of ants in orchards and to heighten their efficiency as predators are the first recorded occurrence of biological control of insects. Specifically, the movement of ants from the countryside into the orchards is an example of augmentative biological control. The use of runways between trees increased the ants' access to prey while keeping them away from potential harm on the orchard floor, so this is also an example of conservation biological control.

In the mid-1880s, southern California's developing citrus industry experienced devastating losses from an introduced pest called cottony cushion scale. Growers tried every available chemical control, even fumigation with hydrogen cyanide, but nothing provided sufficient control; many growers removed their citrus groves because the damage was so serious. After determining that the scale insect was native to Australia and New Zealand, the United States Department of Agriculture sent an entomologist to that area to look for effective natural enemies. The entomologist found a small lady beetle, the vedalia beetle, which he sent to California. It reproduced rapidly in infested citrus groves and brought the cottony cushion scale under complete and lasting control. This was the first highly successful case of controlling a non-native pest by introducing its natural enemies from their native land, a technique now known as classical or importation biological control.

Types of natural enemies

Natural enemies of insects include predators, parasitic insects, nematodes, and pathogens. Successful application of all forms of biological control requires familiarity with these natural enemies and their benefits and understanding how they fit into an overall pest management program. Although the types of natural enemies are discussed in more detail in chapters 4 and 6, a brief introduction is necessary here.

Predators may be insects or other insectivorous animals, each of which consumes many insect **prey** during its lifetime. Predators are often large, active, and/or conspicuous in their behavior, and they are therefore more readily recognized than are parasites and pathogens. Familiar predators of insects include lady beetles, praying mantids, spiders, birds, and bats.

Parasites of insects (also called **parasitoids**) are insects that lay their eggs in or on a **host** insect. When the parasite egg hatches, the young parasite larva feeds on the host (the pest) and kills it (figure 1). Usually that one host is sufficient to feed the immature parasite until it becomes an adult. Many parasites are host-specific, meaning they attack only one or at most a few closely related **species** of host. No insect parasites are harmful to humans or other vertebrates. Although very common, they are not well known because of their small size. One of the smallest, *Trichogramma*, is only about the size of the period at the end of this sentence.

Figure 1. The generalized life cycle of a parasitic wasp, as exemplified by an aphid parasite. (A) Wasp lays egg in a host, in this case a young aphid. (B) As host feeds and grows, parasite larva feeds on host and also grows. (C) When parasite larva is full grown, it pupates within the host, which is now dead. (D) The parasite pupa transforms into an adult wasp, which emerges from the host. After mating, the young wasp seeks new hosts to parasitize.

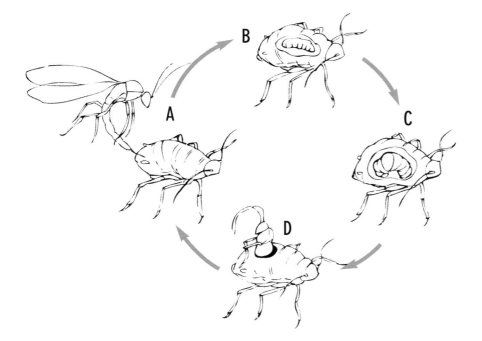

Insects, like other animals, are subject to attack by disease-causing organisms called **pathogens**. Insect pathogens include viruses, bacteria, fungi, and other microorganisms. Disease epidemics among insects are not commonly encountered in nature except when insect populations are large or when environmental conditions favor the growth of the disease organism. Nevertheless, insect pathogens are important in the constant suppression of pest populations. In addition, a growing number of insect pathogens have been formulated into commercial products for biological control of specific pests. The most familiar of these are the different strains of the bacterium *Bacillus thuringiensis*, commonly known as **Bt**, that are available for control of many insects, including various caterpillars, mosquito larvae, and Colorado potato beetle larvae. As is true for insect parasites, most insect pathogens are fairly host-specific and therefore are unlikely to harm **nontarget organisms** such as beneficial insects, humans, livestock, wildlife, or plants.

Insect-parasitic nematodes (also called entomopathogenic nematodes) are small, almost microscopic worms that attack and kill insects that live in moist habitats, especially water and damp soil. These nematodes are not harmful to other animals or plants. A few species are mass-produced and sold for pest control.

The focus of this publication is on discussing and illustrating the important groups of natural enemies of pests and outlining the various approaches to biological control. In addition, to help you best use biological control, we present information on the basics of insect biology, how insects become pests, and how other general approaches to insect management can complement or interfere with biological control.

Examples of biological control

Biological control is the manipulation of living organisms called natural enemies to reduce the abundance of pests. There are many types of natural enemies and several overall approaches to biological control, and these can be used in innumerable combinations. Here are a few examples:

- A greenhouse grower might purchase and release large numbers of a parasitic wasp called *Encarsia formosa* to bring a greenhouse whitefly infestation under control.

- Based on past experience, a greenhouse grower might anticipate spider mite problems and release small numbers of predatory mites early in the plant growth cycle. This way, the predators will already be present when spider mites begin to increase and will be able to prevent an outbreak.

- A market gardener might plant mixed gardens of selected flowering plants throughout the farm. These gardens can provide natural enemies with shelter and food throughout the growing season, keeping them on the farm even when pests are scarce in the crops.

- An apple grower might choose not to apply insecticides for fruit pests during the flight periods of the natural enemies of leafminers. By not killing these natural enemies, the grower will not have to apply another insecticide for leafminers later in the season.

- Scientists search for natural enemies of non-native pests (for example, soybean aphid) in their native land. They import the natural enemies to the United States, carefully screen them for safety and effectiveness, and release them here, in hopes that they will become permanently established and control the pest.

The biology of insects

Knowing basic information about the pests you are trying to control and the natural enemies that attack them puts you in a much better position to achieve the maximum benefit from biological control. You can coordinate releases of natural enemies with a pest's life cycle. You can manage the environment to supply a natural enemy with needed resources and to prevent interference with its beneficial activities. You may be able to manipulate behavioral traits and other life processes of both pests and natural enemies to improve biological control. Finally, knowledge of the relationships among insects is crucial for distinguishing between pest and nonpest species and for recognizing important groups of natural enemies. The more you know about insect biology, the more likely you will be successful with all pest management practices, including biological control.

In this chapter we present some basic but important facts about insect biology that apply to both pest and beneficial insects. In subsequent chapters, we review the natural histories of specific groups of natural enemies and explain the various relationships between pests and natural enemies.

Insect growth and development

Most insects hatch from eggs and go through a mobile but non-flying immature period during which they increase in size. After the insect transforms into the winged, flying, and reproductive adult, it does not increase further in size (figure 2).

Insects have a tough outer skin called the **exoskeleton** (*exo* = outer) or **cuticle**, which protects the insect from adverse environmental conditions. It is usually hardest in insects that are more likely to be exposed to adverse conditions. In order to grow bigger, the immature insect has to shed its exoskeleton and grow a larger one; this process is called **molting**. Most insects molt three to seven times in their lives, and the number of molts is usually constant for each species. For example, alfalfa weevil larvae molt four times, spotted tentiform leafminer larvae molt five times, and European corn borer larvae molt six times. The stage between molts is the **instar**. The larva hatching out of the egg is the first instar; after one molt it is a second instar; and so on.

Figure 2. The upper life cycle shows the common misconception that little insects are smaller versions of adult insects, while the lower life cycle illustrates the correct growth pattern of a beetle. When two similar adult insects are substantially different in size, they are probably different species with different habits.

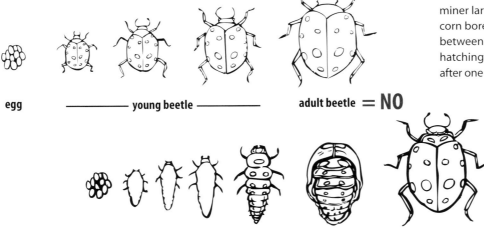

egg ——— young beetle ——— adult beetle = **NO**

egg ——— larva ——— pupa adult beetle = **YES**

When the immature insect has grown to its full size, it transforms into a winged, reproductively active adult through a process called **metamorphosis** (*meta* = change; *morph* = form). Two major categories of insects are distinguished by the complexity of this transformation. For insects with **incomplete** or **simple metamorphosis** (figure 3) the immatures (called **nymphs**) are usually similar to adults in appearance and behavior except that the adults can reproduce and usually have wings.

Insects with **complete metamorphosis** (figure 4) have immature forms called **larvae** (singular, larva), which look very different from the adults, and an intermediate stage called the **pupa** (plural, pupae) during which the transformation in appearance takes place. The larvae are variously known as grubs, maggots, worms, wrigglers, caterpillars, or brood. They may eat completely different foods and live in a different environment than the adults of their species. For example, mosquito larvae live in water and feed on algae and organic matter, whereas the adults are airborne and feed on nectar and blood. When larvae are ready to transform into adults, they enter the inactive pupal stage. The pupa may be encased inside a silken cocoon or other protective structure composed of soil, leaves, debris, or other material.

Examples of incomplete versus complete metamorphosis

Incomplete metamorphosis	Complete metamorphosis
aphids	ants
cockroaches	bees
crickets	beetles
grasshoppers	butterflies
leafhoppers	fleas
lice	flies
plant bugs	gnats
scale insects	lacewings
stink bugs	midges
termites	mosquitoes
thrips	moths
whiteflies	parasitic wasps
	sawflies
	wasps
	weevils

The length of time spent in each of the life stages varies with different species of insects. Mayflies typically are nymphs for several months, but the adults live only a few hours or days. Many moths live less than a week as adults, whereas adult fleas may live 1 or 2 years. For a given species, the duration of each life stage is affected by environmental factors, especially temperature.

Figure 3. A stink bug is an example of an insect with incomplete metamorphosis. After hatching from the egg, the nymph grows, occasionally shedding its skin (molting), until it reaches the winged and reproductive adult stage, after which it no longer grows or molts.

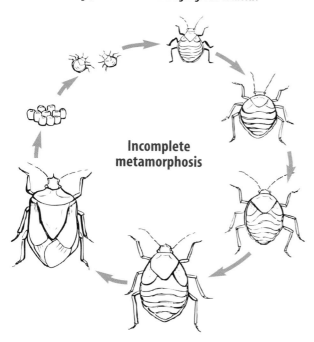

Incomplete metamorphosis

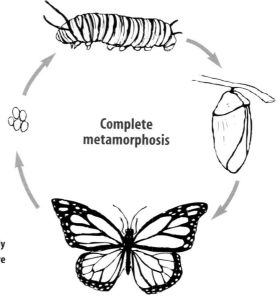

Complete metamorphosis

Figure 4. The monarch butterfly (at right) is an example of an insect with complete metamorphosis. After hatching from the egg, the larva grows, occasionally shedding its skin (molting), until it is fully grown. The larva then molts one more time and transforms into the pupa. The pupa in turn molts and transforms into the adult winged and reproductive stage, after which it no longer grows.

Insects are cold-blooded, meaning their body temperature is very close to that of the surrounding environment. Temperature determines how quickly or slowly an insect grows and develops to its next stage. For example, an alfalfa weevil egg may hatch in 6 days at 75°F, but it will take 20 days to hatch at 55°F. Each insect life stage also has a temperature below which it does not develop at all. At or below this temperature, called the **developmental threshold**, the insect's development ceases until the temperature again rises above the developmental threshold. For the alfalfa weevil egg, this threshold is 48°F. By monitoring environmental temperatures and by knowing an insect's inherent developmental threshold and rate, one can estimate when an insect will develop to a particular stage or event, such as emergence of overwintered adults, egg-laying, or egg hatch. Entomologists have used this type of information, called degree-day analysis, to predict the presence of damaging stages of pests or of stages susceptible to control measures. Degree-day models are useful in biological control because they can guide your use of natural enemies that attack specific life stages of the pest. Your Cooperative Extension Service or crop consultant may be able to provide information for making such predictions for specific pests.

The sequence of life stages from egg to nymph to adult to egg (for insects with incomplete metamorphosis) or from egg to larva to pupa to adult to egg (for those with complete metamorphosis) is called an insect's **life cycle**. It may take days, weeks, months, or even years to complete, depending on the insect species and environmental conditions. Knowing the life cycles of both pests and natural enemies is essential in biological control. Some stages in the life cycle of a pest may be more exposed and/or vulnerable than other stages that may be hidden or protected. Also, the life cycle of a promising natural enemy may be incompatible with the pest you are trying to control. For example, the *Trichogramma* wasp sometimes used to control codling moth overwinters in the egg of its host, but codling moth overwinters as a larva, not as

an egg. If *Trichogramma* cannot overwinter in a different kind of host egg, it must be reintroduced into the orchard each spring.

The number of life cycles, or generations, an insect undergoes per year depends on the species, location, and environmental factors. The number may be fixed (as with the Japanese beetle and corn rootworm beetles, which have one generation per year) or it may vary with factors such as temperature, latitude, or food quality (as with mites and most aphid species).

Each species of insect has a typical pattern of winter hibernation, migration, and/or summer dormancy. In the upper Midwest, most insects undergo a dormant **hibernation** (called diapause) during the winter. Insects may overwinter in any life stage, but this is usually the same for all members of a given species. Many insects, like potato leafhoppers, flower thrips, corn earworm, and some cutworms, are killed off by northern winters and migrate northward each spring and summer from southern areas. Alfalfa weevil, certain lady beetles, and some other insects also undergo a prolonged dormancy during the heat of summer. The widespread use of corn-soybean rotations across the Midwest has selected for an extended dormancy in some populations of northern and western corn rootworms, allowing them to remain dormant in the soil through the soybean part of the rotation.

Reproduction

Most insects reproduce by sexual reproduction, which involves mating and transfer of sperm from the male to the female for egg fertilization. Some insects can reproduce without mating, such as aphids and some insect-parasitic wasps. These insects produce only female offspring for many generations; they produce males only under particular environmental conditions. In a few species, males are entirely unknown.

Most insect populations are composed of approximately half females and half males. However, in some species or under certain environmental conditions, the sex ratio can change substantially. For biological control, the ideal population of natural enemies should contain either equal numbers of both sexes or more females. Producers of natural enemies should pay close attention to the sex ratio of the natural enemies they sell as part of their quality control program.

The process of egg-laying is called **oviposition**. A female places her eggs with her **ovipositor** (figure 5), a structure at the tip of the **abdomen**. Eggs may be laid singly or in clusters. Many insects lay eggs on the surface of plant or animal tissue, soil, or other objects. In other species the ovipositor is modified for inserting the egg inside a host plant, animal, or other object. For example, some wasps that parasitize wood borer larvae can drill through solid wood with their ovipositors to parasitize their hosts.

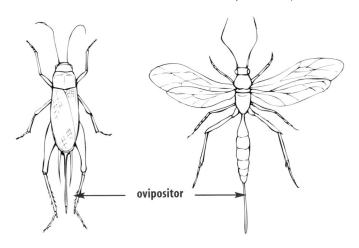

Figure 5. A cricket (left) and an insect-parasitic wasp (right), each with an obvious ovipositor. The cricket inserts its eggs in soil and the parasitic wasp "stings" its host insect, laying its egg inside the host.

The insect body

Insects have segmented bodies with three basic body regions: head, thorax, and abdomen (figure 6). The **head** bears the eyes, antennae, and mouthparts and contains the brain. The **thorax** is directly behind the head and bears three pairs of jointed walking legs and the wings, if present. The abdomen is usually as long as, or longer than, the head and thorax combined. It contains the gut and reproductive organs. The three basic body regions of immature insects can be hard to distinguish (figures 7 and 8). Immatures will be described as needed in chapter 6 and elsewhere.

Adult insects have mouthparts that reflect their feeding habits (figure 9). Beetles, bees, ants, grasshoppers, lacewings, and dragonflies have chewing mouthparts and feed by chewing solid foods. Insects with piercing and sucking mouthparts have a sharp, hollow beak-like structure used to penetrate plant or animal tissue and suck up fluid. These are found in aphids, scale insects, leafhoppers, plant bugs, mosquitoes, and all of the predatory true bugs. Lapping and sponging mouthparts are found primarily in some flies, including the house fly. Butterflies and moths have a coiled proboscis that they uncoil to siphon liquid such as nectar from flowers. Some adult insects, such as mayflies and some moths, do not feed at all and have nonfunctional mouthparts.

Mouthparts of immature insects may be similar to those of the adults, or they may be very different or much simplified, particularly for insects with complete metamorphosis. Caterpillars and some beetle grubs have well-developed chewing mouthparts, but other beetle larvae, many fly maggots, and larvae of parasitic wasps have reduced mouthparts for absorbing liquid foods. Pupae do not feed and therefore do not have functional mouthparts.

Figure 6. All insects have three body regions: head, thorax, and abdomen.

Figure 8. This caterpillar shows the basic body form of a larval insect. Certain groups of larvae, especially caterpillars and sawfly larvae, have fleshy legs, called prolegs, on the abdomen, in addition to six jointed legs on the thorax.

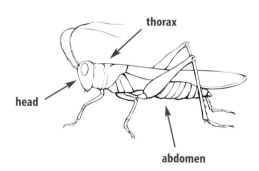

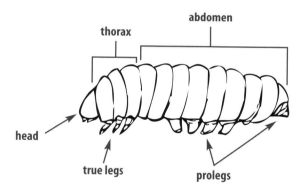

Figure 7. In some larval or highly modified insects, such as this onion maggot, the three standard body regions may be difficult to identify, and other structures such as legs and mouthparts may be modified, reduced, or absent.

Figure 9. Chewing mouthparts of an ant (left) and sucking mouthparts of a cicada (right).

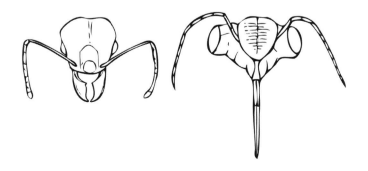

Behavior

Behavior, in its simplest sense, is the response of an organism to some sort of situation in its life. The vigorous wiggling action of a caterpillar when it is disturbed is an attempt to try to escape predation, as is the feigning of death by many beetles. The attraction of moths and other night-flying insects to lights is another type of behavioral response. Plant-feeding pests are usually attracted to their host plants by a combination of stimuli that includes odors and visual characteristics such as color, height, and shape. Similarly, a natural enemy may be attracted first to the specific plant used by its host and then directly to the host by odors, sounds, colors, shapes, or vibrations associated with the host. Natural enemies often have highly specialized behavioral responses to specific stimuli given off by their hosts, which make them very well adapted to finding their specific host species.

Insects have an external chemical communication system for sending messages to other individuals of the same species. Insects release behavioral chemicals called **pheromones**, which serve such functions as signaling danger to nestmates or pointing out a plentiful food supply. Females produce sex pheromones to help males find them. Insect pheromones can be exploited in many ways to help control pests. Traps baited with a pest's sex pheromones can monitor the presence of that pest and the timing of mating. For example, apple growers use this technique to detect the presence and seasonal occurrence of pests such as codling moth, redbanded leafroller, obliquebanded leafroller, and San Jose scale; corn growers can do the same for European corn borer, corn earworm, and armyworm. Sex pheromones are occasionally used as a pest control method. For example, slow-release dispensers of codling moth synthetic sex pheromone deployed throughout an apple orchard confuses male moths sufficiently to disrupt mating, thereby significantly reducing the codling moth population within the orchard.

You also can manipulate the behaviors of natural enemies for biological control. For example, many adult predators and parasites require high-energy foods for moving about to find mates or prey and to mature and deposit their eggs. They often get this nourishment from flower nectar and pollen, the juices of damaged fruit, or the sweet **honeydew** excreted by sap-sucking insects such as aphids. In each case, they are behaviorally attracted to the source of food by odors or other cues. If these foods are not readily available, the natural enemy is likely to leave the area, resulting in poor biological control. To encourage natural enemies to stay in the area, you can provide these necessary resources by manipulating the habitat or by applying artificial foods with appropriate behavioral cues. Chapter 8 contains more information on manipulating natural enemy behavior to improve biological control.

Insect ecology

Like other living things, insects flourish in the correct environment and become weak or diseased, reproduce poorly, or die if their environment is unfavorable. All insects, including pests and natural enemies, need food, water, proper humidity, shelter, mating and oviposition sites, overwintering sites, protection from toxic materials, and protection from their natural enemies. In addition to the needs of individuals, insect populations also have specific requirements for survival. The interactions of insects both within their own populations and between those populations and other animals and plants are often quite complex, and these relationships help determine the size of pest and natural enemy populations and the extent of damage caused by the pests.

Understanding the factors that limit a pest species or favor a natural enemy contributes to better biological control. When considering pest populations, it is most useful to understand the factors that limit population growth, including responses to weather, food availability, overcrowding, and natural enemies. Considerations for increasing and maintaining natural enemy populations include the abundance of hosts, alternate food sources, diversity and structure of habitats, and absence of detrimental factors such as toxic pesticide residues.

Classification

In an effort to make some order out of the incredible diversity of organisms in the world, scientists undertake to arrange them into related categories and give them names that are standardized worldwide. Some familiarity with classification is essential for those with interests in biological control.

Classification is based on similarities that are shared by groups of related organisms. All the animals that have backbones are grouped together because of this characteristic, but within this diverse group, it is easy to distinguish groups of more closely related animals, such as fish, reptiles, birds, and mammals. Insects are classified in a similar fashion (beetles, flies, moths and butterflies, wasps, etc.). Each of these groups can be further subdivided using more specific criteria to define successively smaller groups until the smallest, most specific category, the species, has been identified. The species category is unique because it defines a group of individuals that can mate with one another and produce viable offspring; successful reproduction almost never occurs between individuals of different species. A **genus** is a group of similar, closely related species; for example, *Trichogramma minutum* and *Trichogramma pretiosum* are two distinct wasp species that parasitize caterpillar eggs, but they differ slightly in their biological characteristics.

Table 1. Classification of European corn borer and three of its natural enemies.

Category	Pest: European corn borer	Predator	Predator	Parasitoid
Kingdom	Animalia (all animals)	Animalia	Animalia	Animalia
Phylum	Arthropoda (the arthropods)	Arthropoda	Arthropoda	Arthropoda
Class	Hexapoda	Hexapoda	Hexapoda	Hexapoda
Subclass	Insecta (the insects)	Insecta	Insecta	Insecta
Order	Lepidoptera (moths and butterflies)	Coleoptera (beetles)	Coleoptera (beetles)	Hymenoptera (bees and wasps)
Family	Pyralidae (snout moths)	Coccinellidae (lady beetles)	Coccinellidae (lady beetles)	Braconidae (braconid wasps)
Genus	*Ostrinia*	*Coleomegilla*	*Harmonia*	*Macrocentrus*
Species	*nubilalis*	*maculata*	*axyridis*	*grandii*
Scientific name	*Ostrinia nubilalis*	*Coleomegilla maculata*	*Harmonia axyridis*	*Macrocentrus grandii*
Common name	European corn borer	pink lady beetle	multicolored Asian lady beetle	no common name

The scientific name (= Latin name) of an organism consists of two parts, the name of the genus and the name of the species, the first capitalized and both in italics (or underlined). Table 1 illustrates the classifications of a common pest species, the European corn borer, and three of its natural enemies. The scientific name is always latinized and is written exactly the same in all languages worldwide. Common names, on the other hand, can be confusing, since one species may have several common names. For example, many people refer to *Harmonia axyridis*, the multicolored Asian lady beetle, as the fake lady beetle, the Halloween beetle, or the Japanese beetle.

In other cases, the same common name may be used for different species in different areas. For example, the common name Japanese beetle is also used for *Popillia japonica*, a scarab beetle that is a serious pest of turf and other plants. Similarly, spider mites are sometimes called red spiders. Although scientific names are preferred, the Entomological Society of America has approved frequently used common names for certain highly recognizable insects. We shall use these common names where possible in this publication. However, most natural enemies do not have such common names, making it necessary to use scientific names, such as *Encarsia formosa* and *Bacillus thuringiensis*. For the same reasons, most producers of commercially available natural enemies also use scientific names.

Insects as pests

There are approximately 90,000 named species of insects known in North America, and over 750,000 worldwide. Worldwide, entomologists estimate that there may be 4–6 million species of insects, meaning that most species have yet to be discovered, identified, named, and studied. Despite this incredible diversity, only a small percentage of insect species cause injury to humans and domestic animals or do damage to our crops or possessions. The remainder are either beneficial to humans or the environment, or have no significant recognizable impact, either positive or negative.

A pest is any organism—microbe, plant, or animal—that is a nuisance or causes injury to humans, domestic animals, crops, stored products, buildings, or possessions. Agricultural pests include the microorganisms that cause plant diseases (viruses, phytoplasmas, bacteria, and fungi), plant-parasitic nematodes, weeds, certain vertebrate animals (rodents, birds, deer), certain **arthropods** (insects, mites, and millipedes), mollusks (snails and slugs), and occasionally other organisms. Although biological controls have been used against pests in all of these groups, this publication deals only with the control of insects and mites.

Types of crop pests

Not all types of pests are equally suitable targets for biological control. It is helpful to understand the different ways that insects cause damage in order to understand the strengths and limitations of biological control against various categories of pests. In particular, the suitability of a pest as a target for biological control is influenced by how it feeds, where it feeds, the type and severity of the damage it causes, and its ability to transmit the microorganisms that cause plant diseases.

Method of feeding

A pest's method of feeding is important to biological control for two reasons. First, it determines the type and severity of damage the pest causes. Second, some microorganisms that are used as natural enemies must be ingested to be effective, and these are not effective against sucking insects.

Crop pests feed in two primary ways: chewing and sucking. Sucking insects have mouthparts for piercing plant tissues and sucking out plant sap. Because they do not chew, they do not leave noticeable holes in the plant tissue. Large populations of these insects can remove enough moisture and nutrients from a plant to reduce its vigor. Many inject a saliva into the plant during feeding, which can distort normal growth (some aphids) or cause a toxic burn to the foliage (four-lined plant bugs, potato leafhoppers). Additionally, many sucking insects transmit microorganisms that cause plant diseases. Sucking insects include aphids, leafhoppers, plant bugs, stink bugs, mealybugs, whiteflies, and scale insects. Spider mites cause plant damage similar to that of sucking insects, although their actual feeding mechanism differs somewhat.

Chewing insects have mandibles equivalent to our jaws and teeth, which are used for grabbing, tearing, and chewing food. Chewing directly removes plant tissue, resulting in holes in foliage or fruit, jagged leaf edges, tunnels through leaves, stems, wood, or fruit, or the loss of roots. Caterpillars, grasshoppers, beetles and their larvae (grubs), and sawfly larvae all feed by chewing.

Feeding location

Different insects feed in different locations on the plant. Leaves, stems, branches, bark, buds, flowers, fruit, seeds, roots, tubers, and bulbs can all be attacked. All insects specialize somewhat in their feeding. For example, a stem-boring caterpillar is not usually also a leaf feeder.

Pests that feed in more protected locations may only be susceptible to natural enemies during just part of their life cycle, or they may be attacked by fewer types of specialized natural enemies. For example, codling moth eggs are laid on the surface of apple leaves or fruit, and newly hatched larvae soon tunnel into the core of the fruit where they feed in relative safety. Biological control of codling moth is fairly low, because natural enemies must attack during the brief interval between when the egg is laid and when the larva enters the fruit. In contrast, foliage feeding insects are exposed to natural enemies for much more of their life cycle and may experience much greater levels of biological control.

Beneficial activities of insects

The majority of insects are not pests. In fact, many are of benefit to humans and our environment.

Benefits to humans

Pollination of crops
numerous types of bees, moths, beetles, flies, and other insects

Sources of nutrition
various types of insects are significant dietary components of many peoples worldwide; insects also provide nutrition for fish, poultry, and other human food sources

Insect products
honey, beeswax, silk, shellac, and certain dyes are all produced by insects

Protection, of food, fiber, timber and human health
various predatory and parasitic insects are important natural enemies of pest species (see chapter 6)

Human health
sterile surgical maggots are used for cleaning certain types of wounds; certain pharmaceutical chemicals are derived from insects; some vaccines are produced using insect cell lines

Benefits to the environment

Recycling of nutrients
insects are extremely important in the breakdown of dead plant and animal material and the recycling of nutrients from these materials

Pollination of wild plants
many trees, shrubs, wildflowers, and other plants require insect pollination in order to develop seeds and reproduce

Regulation of plant and animal species
the environment is a complex system of checks and balances; insects are extremely important in helping to regulate the numbers of other organisms

Providing food for other animals
insects are important dietary components for many types of animals, including other insects, fish, amphibians and reptiles, birds, and mammals

Direct versus indirect damage

Many people, when thinking of crop pests, imagine wormy apples or cabbages full of holes. In these cases, the apple maggot and the imported cabbageworm are feeding on and directly damaging the harvested part of the plant. This type of pest is called a **direct pest**; other examples include alfalfa weevil, corn earworm, cabbage looper, spinach leafminer, codling moth, and blueberry maggot. In contrast, **indirect pests** attack some part of the plant other than that which is to be harvested. By feeding on roots, stems, or leaves, they can sometimes sufficiently stress a plant to reduce crop yield or quality. Rootworms, cutworms, spider mites, spotted tentiform leafminer on apple, and Colorado potato beetle are all examples of indirect pests.

Sometimes the status of a given insect as a direct or indirect pest varies with its life stage or the type or stage of development of the crop. The adult mint flea beetle is a direct pest that chews holes in mint leaves, but its larval stage is an indirect pest that feeds on the root system. Larvae of green fruitworms feed on the foliage of fruit trees before and during bloom (causing indirect damage), but they can feed as direct pests after fruit set. Tobacco hornworms are direct pests of tobacco because the leaf is the crop, but this same insect is an indirect pest when it feeds on tomato leaves.

Plants tend to tolerate moderate amounts of leaf damage without loss of quality or quantity of harvest, and biological controls work very well against indirect pests that are foliage feeders. However, stem borers, which are also indirect pests, can weaken or kill the entire plant, necessitating very effective biological control. Direct pests may be difficult to manage with biological control, depending on the effectiveness of available natural enemies and the level of tolerance for slight direct damage.

Severity and regularity of damage

Not all pests are problems to the same degree. The severity of a pest depends on the particular situation, including the insect itself, the location, and environmental factors.

The most severe and regularly occurring pests are called **primary** or **key pests**. For key pests like plum curculio on apples, cucumber beetles on cucurbits, and tarnished plant bugs in strawberries, natural controls seldom provide adequate control. Without human intervention, they would cause significant crop losses or even make growing the crop completely impractical.

Secondary or **occasional pests** usually occur in association with the crop, but they are generally prevented from causing economic injury either by natural controls or by control measures taken against primary pests; examples include leafrollers in apples, diamondback moth in brassicas (cole crops), and flea beetles in potatoes.

Potential pests, such as apple and spirea aphids in apples and corn leaf aphids in corn, occur in association with the crop plant, but under normal conditions they are not abundant enough to cause damage.

The severity, or status, of any pest can increase or decrease if crop management practices, such as pesticide use, change significantly. Such changes in status are usually long term. For example, since the 1960s, planting wheat varieties resistant to Hessian

Isn't "native" good and non-native bad?

Generally speaking, a native species is one that has been present in a given region for millenia. It has been there long enough for it to have fully adapted to the local environment, and for the other species in the region to have adapted to its presence. In other words, it is a fully integrated member of the biological community. Centuries of human activity have brought many non-native species to this continent, sometimes on purpose, but often not. Most of our crops and domesticated animals are not native to North America and were brought here intentionally. However, many of our most serious weeds, plant pathogens, and insect pests are also non-native, and most arrived here as accidental introductions. These latter organisms are serious pests, at least in part, because they arrived here without the specialist natural enemies that keep them under control in their native environments. Classical biological control, or the importation of natural enemies, is an attempt to restore some regulation of these pests by introducing their specialist natural enemies.

Critics of biological control often claim that these non-native natural enemies could go on to become pests in their own right. To support their claims, they typically point to natural enemy introductions that happened long ago, often on tropical or subtropical islands, and often involving birds, snakes, toads, or other vertebrate natural enemies. There is no question that these introductions were mistakes that harmed nontarget species. However, approximately 2,000 insect species have been used in classical biological control programs targeting insect pests around the world, and only about 10 of these species have caused population-level effects on nontarget organisms. In the United States and many other countries, the importation process is more tightly regulated than ever before. Non-native natural enemies must be screened extensively for possible nontarget impacts before they can be released into the environment, reducing the risk of such introductions even further. When done properly, introduction of non-native natural enemies may be the safest, most sustainable way to control many of our most troublesome pests.

In other sources, you may see the terms "introduced," "alien," "exotic," or "non-indigenous." These are all synonyms for non-native, though "introduced" may be used to indicate a non-native species that was established intentionally. For consistency, we chose to use the term non-native throughout this document, but you should be aware that these other terms are in common usage.

fly and postponing planting until after predicted "fly-free" dates have significantly reduced the status of this pest. Similarly, switching to no-till corn can increase the severity of cutworm problems, despite generally higher numbers of natural enemies.

Potential pests and occasional pests often have effective natural enemies that keep them from causing damage. Even many key pests may have effective natural enemies that have been rendered ineffective for various reasons, such as heavy use of broad-spectrum pesticides. Other key pests may lack highly effective natural enemies; these may be excellent targets for new classical biological control programs (see chapter 7).

Plant disease transmission

Many sucking insects, especially aphids and leafhoppers, transmit plant pathogens; this is often the most serious type of damage that these insects cause. Transmission of plant pathogens is less widespread among chewing insects, but bean leaf beetle, corn flea beetle, and cucumber beetles all commonly transmit important plant pathogens. The speed and mechanism of transmission vary among insects. Those insect **vectors** that rapidly transmit a serious disease at the onset of feeding generally are not good candidates for biological control because they transmit the pathogen before natural enemies kill them. Even small populations of such vectors are capable of infecting many plants. For example, aster leafhopper can transmit a disease called aster yellows to carrots, celery, lettuce, and other crops. Although fewer than 5% of leafhoppers carry the organism that causes the disease, pesticide treatment has been recommended for highly susceptible crops when a single aster leafhopper is captured in 20 sweeps of a sweep net.

Other types of pests

Although biological control has been used primarily against plant pests, most insects have natural enemies, and pests of resources other than plants can also be targets for biological control.

Mosquito larvae and the larvae of stable flies and house flies can be important pests of human and animal health. Such insects can be managed with natural enemies in conjunction with other control approaches such as sanitation. Many pests of grain and other stored products are also attacked by natural enemies which can be manipulated or augmented to reduce pest populations. Ants and cockroaches have also been targets of biological control.

How insects become pests

Different species of insects have achieved pest status in different ways. By knowing the various reasons why insects are pests, you can better understand why some are more or less suitable as targets for biological control.

Non-native pests

Many of our serious insect pests are not native to North America but were accidentally introduced from other parts of the world, especially from the 1500s through the 1800s (table 2). During this early period, people were unaware that plant material might harbor pests. No laws restricted the movement of infested plant materials such as live plants, fruits, vegetables, seeds, grains, and tubers. Modern plant-health quarantine regulations help keep hundreds of serious pests out of our area, although even today new pests are occasionally introduced, such as Russian wheat aphid and Asian tiger mosquito, which were detected in the 1980s, pine shoot beetle and Asian longhorn beetle in the 1990s, soybean aphid in 2000, and emerald ash borer in 2002.

Table 2. Examples of non-native pests for specific crops in the North Central United States.[a]

Pest	Origin	Pest	Origin
Field and forage crops		**Forest trees and landscape plants**	
alfalfa weevil	Europe	Asian longhorn beetle	Asia
cereal leaf beetle	Europe	balsam woolly adelgid	Europe
clover leaf weevil	southern Europe	beech scale	Europe
clover root weevil	Europe	birch leafminer	Europe
European corn borer	Europe	black vine weevil	Europe
greenbug	Europe	Comstock mealybug	Asia
Hessian fly	Europe	elm leaf beetle	Europe
Russian wheat aphid	Asia	emerald ash borer	China
seedcorn maggot	Europe	elm leafminer	Europe
southwestern corn borer	Mexico	euonymus scale	Asia
soybean aphid	Asia	European elm scale	Europe
spotted alfalfa aphid	Middle East	European pine sawfly	Europe
sweet clover weevil	Europe	European pine shoot moth	Europe
		European spruce sawfly	Europe
Vegetable crops		gypsy moth	Europe
asparagus beetle	Europe	introduced pine sawfly	Europe
beet armyworm	Asia	larch casebearer	Europe
cabbage maggot	Europe	larch sawfly	Europe and Asia
carrot rust fly	Europe	maple petiole borer	Europe
Colorado potato beetle	Mexico	mimosa webworm	China
diamondback moth	Europe	pine bark aphid	Europe
imported cabbageworm	Europe	rhododendron whitefly	Asia
Mexican bean beetle	Mexico	smaller European elm bark beetle	Europe
potato tuberworm	South America		
spinach leafminer	Europe	wooly hemlock adelgid	Asia
spotted asparagus beetle	Europe		
		Greenhouse crops[b]	
Fruit crops		bulb mite	Europe
codling moth	southeastern Europe	chrysanthemum gall midge	Europe
		citrus mealybug	China
European red mite	Europe	narcissus bulb fly	Europe
green peach aphid	Europe		
imported currantworm	Europe	**Stored products**	
Japanese beetle	Japan	Angoumois grain moth	unknown[c]
oriental fruit moth	Asia	cigarette beetle	unknown[c]
oystershell scale	Europe and Asia	confused flour beetle	unknown[c]
pear leaf blister mite	Europe	granary weevil	unknown[c]
pear psylla	Europe	Indian meal moth	unknown[c]
San Jose scale	China	Mediterranean flour moth	unknown[c]
shothole borer	Europe	rice weevil	unknown[c]

[a] This list is not all-inclusive but demonstrates the diversity of serious pests of agriculture, horticulture, and forestry that have been introduced and established in the North Central United States by human activities, mostly accidentally. The list does not include nuisance pests, household and structural pests, and pests of human and animal health.

[b] Many additional greenhouse pests are distributed virtually worldwide and the origin of most is unknown. However, many are probably not of North American origin.

[c] Pests of stored products have been distributed worldwide for hundreds of years, making it difficult to determine their origins. Many probably originated in the Mediterranean region, the Middle East, and the Indian subcontinent.

The impact of non-native pests and the potential for biological control of these insects are affected by the interaction between each pest species and its natural enemies. Close relationships develop between an insect and its natural enemies as they evolve. Some natural enemies will only attack a single species of host or prey (pest). These highly specialized natural enemies are very well synchronized with their hosts or prey and have well-developed behavioral patterns that allow them to find and kill their hosts or prey, even when they are uncommon. These traits are often considered ideal for successful biological control, but are not always necessary.

When a pest is accidentally introduced into a new area, some of its **specialist natural enemies** also may become established in the new area. This situation is termed fortuitous biological control. However, most non-native pests arrive here unaccompanied by their specialist natural enemies. Native **generalist natural enemies** may provide some natural control of this new pest, but usually they are not as effective as the specialists in the pest's native locale. Importation of natural enemies is a deliberate attempt to establish these specialists for control of previously introduced non-native pests (see chapter 7).

Non-native crops

The reverse of the introduction of a non-native pest into an area with suitable crop plants is the introduction of a non-native crop into an area where a native plant-feeding insect can feed on it and becomes a pest. While not as common as the previous situation, it does occur and can have significant implications for biological control.

Most plants have defensive mechanisms (often chemical toxins) that keep insects from eating them. This is one reason why every species of plant-feeding insect is not capable of feeding on every species of plant. When a non-native crop plant is introduced into a new area, it may lack the necessary mechanisms to defend itself from attack by all the species of plant-feeding insects in the new locale; certain insects may be able to feed on this newly introduced plant species, and thereby become pests. Additionally, many natural enemies locate their insect prey by first locating the plant host of the prey, often by sensing odors given off by the plant. Native natural enemies may not recognize the odors produced by the newly introduced crop plant and, therefore, cannot control pests feeding on that plant.

Alfalfa caterpillar and apple maggot are two examples of native insects that fed on native plant species but rapidly adapted to the introduced crops that they now damage.

Monocultures

A **monoculture** is a large contiguous planting of one crop with little or no additional cultivated or wild vegetation. In large-scale farming, monocultures are easier to plant, manage, and harvest than smaller plots. However, monocultures provide pest insects with an abundance of food. The pest expends little time or energy finding food or a place to lay its eggs. Pest populations can rapidly increase in such situations.

When different plant species grow in close proximity (a **polyculture**), many different species of plant-feeding insects and natural enemies are present in the area. This increases the likelihood that one or more of the natural enemies will help suppress a growing pest population. Because monocultures have few species of plants, they tend to discourage the development of a healthy mixture of generalist and specialist natural enemies. When dealing with the complexities of biological interactions, however, there are no absolutes, and in some cases pest numbers are actually higher in polycultures than in monocultures.

Interference with natural enemy activity

Many agricultural and pest management practices reduce natural enemy numbers or interfere with their ability to control pests. Foremost in this category is the use of broad-spectrum insecticides, to which most natural enemies are highly susceptible.

By the early 1950s, chemical pest control was accepted as an easy, effective, and economical practice. Pesticide use dramatically increased yields of high-quality food and there was little evidence, at that time, of potential problems. Crops with serious primary pests or with multiple primary and secondary pests were routinely treated chemically. Although these treatments eliminated the primary pests, secondary and potential pests began to cause increasing amounts of damage. Investigation revealed that the pesticides were eliminating natural enemies along with the key pests, allowing occasional or potential pests to increase to damaging levels. This situation is called a **secondary pest outbreak**.

It was also found that **pest resurgence** often occurred following a pesticide application: a pest species with multiple generations per year would rapidly rebound to damaging levels. With less than 100% control from a pesticide application, and with migration of the pest into the treated field from surrounding areas, the pest population would increase rapidly in the absence of natural enemies.

Because of secondary pest outbreaks and pest resurgence—both attributable, in part, to natural enemy elimination by use of pesticides—growers often increased pesticide concentration and/or the number of sprays, which further compounded the situation.

Many growers now practice **integrated pest management** (**IPM**). Ideally, IPM reduces insecticide use through the use of routine crop monitoring and pest scouting, combined with a knowledge of multiple pest control approaches and the levels of pest abundance at which damage becomes economically significant. When pesticide use is necessary, the most selective chemicals and selective application methods available are chosen to protect natural enemies (see chapter 8). Reduced use of broad-spectrum pesticides makes natural enemy survival more likely, and pest resurgence and secondary pest outbreaks are less likely.

Broad-spectrum insecticides are not the only factors that disrupt natural enemies. Anything that interferes with a natural enemy's requirements for food, water, shelter, and mates will reduce its effectiveness. For example, the adults of many predatory and parasitic insects must feed on some carbohydrate source for energy and egg production. Such foods can be found in damaged or decaying fruit, insect honeydew, and flower nectar and pollen. If appropriate sources of these materials are not locally available, the adult insects will likely fly elsewhere for feeding and production of offspring.

Reduction of pesticide impacts on natural enemies and provision of resources that natural enemies requires are two major emphases in conservation biological control, which is discussed further in chapter 8.

Pesticide resistance

Resistance occurs when individuals within a population are capable of surviving some mortality factor to which other individuals are susceptible. If exposure to the mortality factor continues, only the resistant individuals survive to reproduce, which increases the number of offspring that are also resistant. In agricultural settings, resistance to pesticides is a major concern. The development of pesticide resistance can be slowed or avoided by reducing the frequency of pesticide application or by rotating among different pesticides, especially if the pesticides are in different chemical classes with different modes of action.

In a few documented cases, natural enemies have also developed resistance to certain groups of pesticides. Predatory mites of the family Phytoseiidae, which are important predators of spider mites in orchards and other crops, have naturally developed resistance to organophosphate insecticides in some areas of the country. Such resistance allows the integration of chemical with biological control approaches, because growers can use certain insecticides without adversely affecting the predatory mites. Artificial laboratory selection for resistance in natural enemies, and the subsequent release of these into chemically treated crops, has also been successful in a few limited situations.

How to tell if an insect is a pest

Integrated and sustainable pest management practices require a working knowledge of pests and beneficial insects, especially the recognition and identification of the various insect species present in a crop. Not all insects are pests, but how can you determine what is, and what is not, a pest?

Simple observation may answer the question. Large green larvae chewing holes in cabbage heads are certainly pests. Medium-sized, oval, yellow and black striped beetles defoliating potato plants are definitely the Colorado potato beetle, a serious pest. However, other pest species are much less obvious. Aster leafhoppers can be very important pests even in low numbers because they transmit the pathogen that causes the serious aster yellows disease, but the leafhopper is easily overlooked because it is tiny and moves rapidly.

Sometimes insect feeding may look serious when it really is not of economic concern. A colony of large eastern tent caterpillars in a fruit tree is impressive, but unless they cause significant defoliation in consecutive years, they do very little harm to the tree or the crop.

Remember: many insects in a crop or garden are beneficial or are merely transients with no particular impact on the crop, and many plants can handle moderate defoliation with no loss in vigor or yield.

Many useful resources are available for insect identification. Various state and regional Extension publications contain excellent diagnostic pictures of serious pests and their damage. Lists of Extension publications are available from local county Extension offices or from state Extension publication offices or their websites.

Most libraries and book stores also have reference books and field guides that are helpful in recognizing pest and nonpest insects. No single reference can include all the 90,000 different species of insects found in North America, but field guides usually illustrate the most common species in each major insect group, including groups of pests, natural enemies, pollinators, and incidental insects. Image searches on the internet can also be useful, providing you search using appropriate keywords. General books on pest control also illustrate common pests and natural enemies; some of these are listed in "Additional Reading" at the end of this publication.

Economics of pest management

At its simplest, the economic success of crop production is determined by the difference between the costs of production and the revenue from marketing the crop, which depends not only on quantity and quality of the crop, but also on how it is marketed. Pests and pest management affect both sides of this simple equation. Pests can reduce both quantity and quality of the crop via any of the mechanisms discussed earlier in this chapter, though nearly all crops can sustain some level of pest damage without the loss being economically significant. Pest management practices (see chapter 5) can add considerably to production costs, so it makes sense only to apply a control measure when its cost is less than the incremental value of the portion of the crop protected from damage.

The pest abundance at which the resulting economic injury exactly equals the cost of control is called the **economic injury level** (EIL). When control measures are taken at pest levels below the EIL, then the cost of control exceeds the benefits. On the other hand, if control measures are postponed until pest abundance exceeds the EIL, then some economic injury will have occurred despite the control measures. Because it is usually impossible to apply control measures exactly when a pest population reaches the EIL, control measures are usually initiated at a pest abundance slightly below the EIL, which is called an **economic threshold** (ET). Applying control measures when pests reach the ET allows for economic pest control despite possible time lags in implementing control measures and their taking effect.

You may see **action threshold** (or action level) used as a synonym for economic threshold. However, other people use action threshold as a more general term than ET. Both terms indicate the pest abundance at which control measures should be taken to prevent damage, but ET specifies that the damage is defined economically. But pests can also cause other types of damage. For example, in landscape settings, aesthetic damage can be a more important consideration than economic damage, so the term action threshold might be preferred. Thus, economic thresholds are simply action thresholds for which economic factors are the primary consideration in determining when control measures are needed.

These concepts are simple in principle and are useful in guiding decisions about whether to apply control measures, and if so which ones, and when. However, economic thresholds are dynamic and difficult to determine for a variety of reasons, including (1) accurately predicting the actual effect of each pest on the quantity and/or quality of the crop, (2) determining the actual abundance of each pest, (3) variability in costs of managing the pests, including materials, equipment, labor, and monitoring, (4) differences in effectiveness and rapidity of the selected control measures, and (5) fluctuations in the market value of the crop.

CHAPTER 4

Natural control of pest insects

The size of every population of every species is regulated by natural environmental factors, which differ by location and change through time. Various factors may combine to substantially reduce a population in one location or make it more abundant in another. Pests respond to these environmental factors just as do nonpest organisms. **Natural control** refers to those natural processes, unaided by human involvement, that help suppress the abundance of pests. The relative impact of natural controls in a given location helps determine whether pest numbers are going to be high or low.

Natural control is an important but often overlooked component of pest control. The actions of nature are often subtle and not always apparent. However, if pest managers are to reduce their reliance on pesticides and other preventive forms of pest control, they must become more knowledgeable about natural controls. Specifically, they need to develop competence in the following areas:

- distinguishing pests from nonpest species;

- understanding the life cycle of the pest and the type, severity, and predictability of damage done;

- understanding how pest populations respond to environmental factors such as weather, geography, and soil conditions;

- recognizing the competitors and natural enemies that help naturally control the pest;

- routinely monitoring for pest and natural enemy activity; and

- knowing effective and environmentally safe methods of controlling a pest when natural and biological controls are not enough; such methods should interfere as little as possible with the natural and biological controls of other pest species in the system.

Abiotic natural controls

Many aspects of the physical environment exert a regulatory effect on pest populations. These **abiotic factors**, or nonbiological factors, include weather, topography, geography, and soil conditions.

Weather

Wind, temperature, rainfall, and humidity all affect insect and mite populations. They can either increase or decrease populations, depending on their severity and the physical tolerances and requirements of specific pest and beneficial species. Often, it is a combination of factors that affects a pest. A prolonged gentle rain will have minimal impact on small pests such as aphids and spider mites, whereas a hard rain with driving winds will wash many such pests from the plants and result in significant natural control.

Continental wind patterns can greatly influence the seasonal distribution and, therefore, the pest status of some insects. Many insect species cannot survive the very cold winters of the northern United States and Canada. Some of these overwinter in the southern United States and Mexico, and are blown northward on the prevailing spring winds. Examples are black cutworm, corn earworm, and potato and aster leafhoppers. Each year the wind-flow patterns differ, thereby influencing the ultimate distribution and regional abundance of certain pests.

Moisture levels also affect pest populations. For example, most spider mite species, such as twospotted spider mite, mature and reproduce fastest under warm, dry conditions. Drought conditions result in large and damaging mite populations on many crops. Many aphid species feed on the rapidly growing succulent tips of plant stems; during drought conditions plant growth is reduced and these aphid species are less abundant. For other pests, drought stress may increase the nutritional quality of their host plants, leading to larger pest populations, as seems to be true for soybean aphid. Also, many aphid species that are normally controlled by fungal insect pathogens may actually be more abundant during dry periods that are detrimental to the fungi.

The growth and reproduction of insects are temperature-dependent. While many insect species have only one or two generations per year regardless of temperatures, some species, including many aphids and mites, can complete more generations in warm years than in cooler years. The extra generations can result in substantially greater pest populations.

Topography and geography

Insect populations can become adapted to the local climate and may not survive or flourish elsewhere unless the climate is very similar. Even if locations with a suitable climate exist elsewhere, major topographical features such as mountain ranges, deserts, and oceans greatly restrict the natural movement of insects. More locally, slope and geographic orientation affect microclimate. For example, north-facing slopes are cooler and more humid than south-facing slopes, and low-lying areas are generally moister, cooler, and more subject to frost than hilltops. On an even smaller scale, in areas with mild winters, Russian wheat aphids will continue to feed and develop throughout the winter in fields with rows running east to west, but not in fields with north-to-south rows, due to differences in microclimate. These types of microclimate differences can greatly influence insect populations.

Soil conditions

Physical, chemical, and biological characteristics of soil all have an impact on soil insects. For example, grubs of the rose chafer are much more common in sandy, well-drained soil rather than heavier, damper soils.

Biotic natural controls

Many **biotic factors**, or biological factors, also limit the size of a population. Competition among individuals or between species for food and other resources can limit population growth. However important competition may be, it is usually impossible for us to manipulate it for controlling pest populations. **Host-plant resistance** is another factor, in which the plant's chemical defenses and physical characteristics (e.g., fuzziness) can deter insect feeding. Of chief importance to those interested in biological control are the effects of natural enemies, which are organisms that kill, seriously debilitate, or otherwise interfere with individuals of another species. Virtually all species, pests and non-pests alike, are attacked by one or more species of natural enemies.

Natural enemies of insects and mites fall into four main categories: predators, parasitic insects, insect-parasitic nematodes, and insect pathogens. Within each of these categories, some natural enemies will attack a wide variety of pests, whereas others attack only one or a few species.

Predators

Although birds, bats, rodents, and frogs contribute to the natural regulation of insect populations, the most important predators in natural control of pest insects and mites are other insects and mites. A high percentage of insects are predators, and many are beneficial in crop settings. Predators are usually mobile and they consume many prey (pests) during their lives. Most predators are generalist natural enemies, although a few types are specialized. A few common examples are lady beetles, lacewings (aphidlions), praying mantids, syrphid flies, assassin bugs, minute pirate bugs, spiders, and predatory mites (phytoseiids).

Parasitic insects

Parasitic insects, often simply called parasites or parasitoids, belong primarily to two major groups of insects: the flies and the wasps. The general life cycle (figure 1, page 2) consists of four stages: egg, larva, pupa, and adult. The adult is a free-living (non-parasitic and mobile) insect such as a fly or wasp. It seeks out the host (pest) insect and deposits one or more eggs in, on, or near it. The egg hatches into the larval stage, which is often maggot-like in appearance. The larva feeds attached to or inside of the host insect, consuming and eventually killing it. Once the larva is fully grown, it changes into a pupa which then changes into the adult. A parasite larva is very closely tied to its individual host insect, is generally incapable of much movement on its own, and kills only one host insect during its development. Many parasites are specialist natural enemies.

Most plant-feeding insects (as well as many other types of insects) are attacked by one or more species of parasite. Indeed, there are many thousands of species of parasitic insects in North America. Most are small (some less than 1 mm when fully grown) and nondescript. It is often difficult to distinguish them from other small insects. Parasitic wasps are mostly incapable of stinging and do not attack humans, unlike their larger relatives such as hornets and yellowjackets.

Insect-parasitic nematodes

Nematodes are tiny roundworms that live in moist habitats. Many species parasitize insects, especially insects living in water (such as mosquito larvae) or in or on moist soil (such as root maggots, grubs, cutworms, and crickets). These nematodes search out insects, parasitize them, and then reproduce, resulting in more parasitic nematodes that will kill additional insects. Most nematodes that attack insects are harmless to other animals and do not attack plants.

Insect pathogens

Insects, like other organisms, are subject to infectious diseases caused by microorganisms called insect pathogens. Many insect pathogens are important in natural and biological control. Insect viruses tend to be fairly host specific and often work very rapidly when host population levels are high. Some bacterial pathogens are host specific while others are more general. Fungal pathogens usually require high humidity to be effective and are fairly common in the Midwest, especially during periods of higher moisture. Many **protists** also attack insects, but they often debilitate their host rather than kill it outright. Most insect pathogens are not harmful to plants, humans, or other types of organisms, although a few can be lethal to beneficial insects as well as pests.

Generalists and specialists

The collection of host or prey species attacked by a given natural enemy is known as its host range. Specialist natural enemies normally attack just one or a few species of prey or hosts. Generalist natural enemies, on the other hand, attack many different prey or host species. With these definitions in mind, and knowing that many exceptions exist, we can make the following generalizations.

Most predators are generalists. They may not leave an area when pest populations are low as long as alternate prey or hosts or other foods are available. Because generalists typically lack the ability to home in on isolated pest infestations, and many have a longer life cycle than their prey or hosts, they are more important in slowing the growth of pest populations, rather than reducing them rapidly. However, a diverse community of generalist natural enemies, including spiders, lady beetles, ground beetles, and many others, can maintain pest populations below levels at which economic injury occurs.

Farming practices that provide alternate prey, hosts, and foods can attract and retain a variety of generalist natural enemies, so that they are ready to get to work immediately when pest numbers begin to increase (see chapter 8). For example, straw mulch around potatoes provides shelter for numerous spiders, ground beetles and other predators of Colorado potato beetle, and patches of flowering plants can offer shelter, floral foods, and alternative prey for many natural enemies. Similar practices will also sustain purchased generalist natural enemies, such as lacewings and minute pirate bugs (see chapter 9).

Most parasites and pathogens, on the other hand, are specialists. They depend entirely on one or a few species of hosts or prey and can quickly find them, even in patchy infestations. For this reason, specialists can be very important in preventing pest outbreaks and many successful cases of classical biological control have involved specialist natural enemies. Specialists often have a short life cycle so their populations can grow quickly when hosts or prey are abundant. As a result, specialists are usually more effective than generalists at quickly reducing established pest populations. Releases of purchased natural enemies intended to quickly control a large pest population (inundative releases) are typically most successful with specialist natural enemies. For example, you can control a greenhouse whitefly infestation using mass releases of the specialist parasite *Encarsia formosa*. This is expensive because large numbers of the parasite are needed. However, since *Encarsia* is good at finding isolated patches of whiteflies, you can also release small numbers of this parasite while pest populations are small and rely on them to prevent a pest outbreak.

Specific groups of natural enemies important in natural and biological control are covered in more detail in chapter 6.

Other approaches to insect pest control

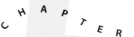
Natural control is extremely important in keeping pest numbers below damaging levels (chapter 4), but populations of insects and mites often escape natural control and develop large, injurious populations. At such times, direct action may be needed to reduce pest numbers and prevent serious damage. Biological control is one of several general approaches to pest management (see chapters 7–10), and involves manipulation of predators, parasites and pathogens to control pests. Other insect control methods can be classified into major categories as follows.

Cultural controls

Many farming practices can reduce pest populations by making their environment less favorable; a few examples follow.

Crop rotation replaces a crop that is susceptible to a serious pest with another crop that is not susceptible, on a rotating basis. For example, corn rootworm larvae can be starved out by following corn with one to two years of a nonhost crop such as soybean, alfalfa, or oats. In some areas, however, populations of northern and western corn rootworms have adapted to consistent corn-soybean rotations by extending their dormancy by an extra year or by laying their overwintering eggs in soybean fields that will likely be planted to corn the following year.

Sanitation refers to keeping the area clean of plants or materials that may harbor pests. Examples include removal of weeds in greenhouses that may harbor mites, aphids, or whiteflies; destruction of crop residues such as corn stubble, squash vines, potato cull piles, or fallen apples that may be overwintering sites for pests; cleaning of farm equipment that can spread pests from field to field; and removal and management of manure that provides breeding sites for flies.

In **strip cropping**, field crops such as alfalfa, soybean, corn, and small grains are planted in parallel strips. This practice creates a more diverse habitat that is favorable for many natural enemies and less favorable for most pest insects.

Insectary plantings are diverse patches of flowering plants that provide natural enemies with resources such as pollen, nectar, shelter, and alternate hosts or prey that might not be available in the main crop. In many cases, **insectary plantings** in or around a cropped area can enhance biological control and reduce pest numbers in the main crop (see chapter 8).

Trap cropping is the provision of a pest insect's preferred food near the crop to be protected. The pests are attracted to the **trap crop**, which is then destroyed along with the insects it harbors. For example, pickleworms will concentrate in squash planted near cucumbers, and the squash plants can be destroyed.

A carefully considered **time of planting** will help avoid some pest problems. For example, planting corn, beans, and peas when few adults of the seed corn maggot are present can significantly reduce damage by the seed-feeding larvae. Cabbage root maggot damage can be controlled in a similar fashion, and Hessian fly has been largely eliminated as a pest of wheat through the adoption of fly-free planting dates.

Host-plant resistance

Some plants have physical and/or chemical adaptations that allow them to repel, tolerate, or even kill pests. Plant breeders attempt to use these characteristics and even improve them to develop crops that are pest resistant. Many varieties of important crops grown today, such as wheat, rice, alfalfa, corn, and apples, are resistant to one or more pests.

Historically, the development of resistant varieties was often tedious and lengthy, requiring many generations of plant breeding. Although such traditional techniques will continue, modern methods of biotechnology already have and will continue to provide pest-resistant crops.

Physical controls

Several practices physically keep insect pests from reaching their hosts. Barriers such as window screens exclude health and nuisance pests from buildings and plant pests from greenhouses. Floating row-covers protect many horticultural crops, and plant collars keep cutworms from attacking plants such as tomatoes. Snails and slugs are reluctant to cross diatomaceous earth or strips of copper. Various types of commercial traps can be used for control, such as pheromone-baited cockroach traps, food-baited traps for ants or yellowjackets, and sticky red spheres with or without food scents for trapping apple maggot flies. Homemade traps can also be effective. For example, codling moth larvae seeking sheltered locations for pupation or overwintering can be trapped under corrugated cardboard bands wrapped around apple trees; the bands can be removed and destroyed before the adult moths emerge. Similarly, gypsy moth caterpillars seek shelter during the day and feed primarily at night. You can greatly reduce gypsy moth numbers and the resulting defoliation of valued trees by wrapping trunks of infested trees with burlap bands and destroying the caterpillars that congregate there each day.

Mechanical controls

Mechanical control methods that directly remove or kill pests can be rapid and effective. Many such techniques are mostly suited for small, acute pest problems, and they are popular with small-scale gardeners. Methods for commercial agriculture have also been developed. An attractive feature of some mechanical controls is that they may have relatively little impact on natural enemies and other nontarget organisms, and can be used in conjunction with biological control in an integrated pest management approach (see below). Cultivation or tillage exposes many soil insects to desiccation or predation by birds. Hand-picking can be used for large or brightly colored foliage feeders such as Colorado potato beetle, Mexican bean beetle, and tomato hornworm. Shaking the plants will dislodge many pests. For example, plum curculio beetles can be removed from fruit trees by diligently banging tree limbs with a padded stick and collecting the adult weevils on a white sheet as they fall out of the trees. A strong spray of water will dislodge aphids and mites from greenhouse, garden, and house plants. Fly swatters and mouse traps also are forms of mechanical control.

Chemical controls

Chemical control is the use of chemicals to kill pests or to inhibit their feeding, mating, or other essential behaviors. The chemicals used in chemical control can be natural products or synthetic materials. Repellents, confusants, and irritants are not usually toxic to insects, but they interfere with their normal behavior and prevent the insects from causing damage. Mothballs and mosquito repellents are familiar examples. Synthetic sex pheromones may confuse insects sufficiently that they are unable to mate and produce offspring; a few such products are commercially available but work best in large commercial plantings where it is less likely that mated females will move into the planting from outside of the treated area.

The following are brief descriptions of some of the major groups of chemicals available to control insect and mite pests. The effect of each of these groups on natural enemies is covered in chapter 8.

Synthetic organic insecticides include a wide range of materials, often based on modifications of petroleum-based molecules. Most of the modern synthetic insecticides currently used in the United States fall into four classes: organophosphates, carbamates, synthetic pyrethroids, and neonicotinoids, which is a relatively new class. Most products in these chemical classes interfere with the pest's nervous system and have a broad range of activity against insects, including natural enemies. Most kill by contact, inhalation, or ingestion. Although some break down fairly rapidly in the environment, most persist for several days or weeks, giving prolonged insect control, but also resulting in prolonged interference with natural enemies.

Inorganic insecticides are derived from mineral elements. Examples in current use include sulfur, hydrated lime, boric acid, and cryolite. Some of these compounds work as stomach poisons, others as repellents, and others by mechanisms that are still poorly understood. In the past, products containing arsenic, such as Paris green and lead arsenate, were widely used as stomach poisons against chewing insects. Because of their hazard to people and animals and their persistence in the environment, arsenic-based compounds are no longer widely used.

Insect growth regulators (IGRs) are synthesized mimics of insect hormones or compounds that interfere with insect development and/or molting. In recent years, IGRs have become an increasingly important group of insecticides. Many of these products are somewhat compatible with biological control.

Microbial insecticides are commercially prepared products that contain active insect-pathogenic microorganisms. **Microbial insecticides** are formulated so that they can be mixed and applied like traditional insecticides. Their use, sometimes called **microbial control,** is generally considered to be a form of biological control and is covered more thoroughly in chapter 9.

Microbially derived insecticides contain toxic compounds derived from microorganisms, but do not contain the actual microorganisms as an active ingredient. Many bacteria and fungi naturally produce chemicals that are toxic to other organisms. In some cases, these microbes can be mass-produced, such as in giant fermentation tanks, and their toxins can be harvested, purified, and formulated into commercial insecticides. Examples include spinosyns and avermectins. Some **microbially derived insecticides** are considered low-risk products that are relatively safe to humans, the environment, and the natural enemies that are important in biological control.

Botanical insecticides are naturally occurring toxic materials derived from plants. Examples include pyrethrum, rotenone, sabadilla, ryania, nicotine, and neem (azadirachtin). Since they are found in nature and are less persistent in the environment, many people consider them to be safer to use than synthetic insecticides. However, "natural" does not mean safe, and some botanical insecticides can be quite hazardous to humans, other mammals, birds, and fish. Most botanicals are nonspecific and highly toxic to many groups of insects, including natural enemies and pollinators. Many degrade quickly in the environment, so they are harmful to natural enemies at the time of application but do not leave persistent toxic residues.

Oils, soaps, and inert dusts all have insecticidal properties and are available commercially, specifically formulated for insect control. Oils kill insects by smothering them. Different insecticidal oils are available for use under different conditions; the proper type has to be chosen or plant injury may occur. Insecticidal soaps can pass through the insect cuticle and poison the insect. They may also act as a detergent that removes some of the waxy waterproofing layer of the cuticle, especially for soft-bodied insects, resulting in loss of body moisture and death by desiccation. Many recipes for homemade soap sprays are available, but not all household soaps are equally insecticidal, and the various additives may cause plant injury; therefore, commercial insecticidal soaps are preferred. Inert dusts, such as diatomaceous earth, boric acid crystals, and silica gel cause insect death in part by interfering with the waterproofing waxy layer of the insect cuticle. Kaolin clay is applied as a particle film that serves as a physical barrier or deterrent to pest landing, feeding, or oviposition on coated plant surfaces. Oils, inert dusts, and soaps generally have little residual activity and therefore may be less detrimental to natural enemies than conventional chemical insecticides.

Acaricides or miticides are synthetic organic or microbially derived pesticides that are primarily effective against mites. They work via a variety of mechanisms, most of which have little to no effect on insects. Most miticides fit well into integrated pest management programs because of relatively low impact on natural enemies, though some are highly toxic to predatory mites.

Chemical controls, particularly synthetic organic insecticides, have been developed for nearly every insect pest. They are widely used in industrialized nations for several reasons: they are highly effective, one product often controls several different pests, there is relatively low cost for product or labor (but relatively high investment in equipment), and their effects generally are predictable and reliable. Chemical insecticides have allowed management of larger acreages by fewer individuals because of the reduced labor needed for physical and mechanical controls. Besides their use in agriculture, chemical insecticides have been very important in the battle against disease-carrying insects, such as the mosquitoes that carry malaria.

Historically, most chemical insecticides have been broad-spectrum and their use has resulted in significant nontarget impacts. There are several disadvantages to such products: most have biological activity against many forms of life and therefore can harm nontarget organisms; for the same reason, they present various levels of hazard to humans, especially pesticide applicators and other farm workers; most are highly toxic to beneficial insects, such as pollinators and predatory and parasitic natural enemies; and resistance to insecticides can develop in both target and nontarget insects, sometimes very rapidly.

There is, however, a trend toward the development of more selective pesticides, but each of these will have to be evaluated for compatibility with biological control on a case-by-case basis. It is difficult to rank the relative toxicity of pesticides because the effects of a pesticide can differ by organism as well as by the life stage of the organism, but table 3 on the following page suggests that at least some pesticides may be compatible with biological control.

Integrated pest management (IPM)

As initially proposed, integrated pest management combines all effective, economical, and environmentally sound pest control methods into a single, flexible approach to managing pests. It is neither possible nor economically feasible to eliminate all pests, so IPM programs aim to keep pest populations below economically damaging levels.

Practitioners of IPM need to recognize the signs and effects of abiotic and biotic natural controls and understand the importance of both. When human intervention is necessary to control pests, they ought to choose the least disruptive practices, such as host-plant resistance, biological control, and cultural control, because these practices are least likely to disrupt natural control; they are also the most sustainable practices. Highly disruptive or environmentally damaging practices should be used only as a last resort. Chemical pesticides, especially those with broad-spectrum activity, should be used only when necessary, based on frequent and routine monitoring of pest populations. Natural enemy populations should also be monitored so that their impact on pests can be determined (see chapter 10). When pesticides are necessary, preference should be given to products that are least detrimental to natural enemies.

Integrated pest management is a dynamic and evolving practice. Specific management strategies will vary from crop to crop, location to location, and year to year, based on changes in pest populations and their natural controls. As specific new methods or technologies are developed, these too can be incorporated into the program as appropriate. Effective pest managers will be as knowledgeable as possible about the pests, the natural enemies, and all available control options. For more information on IPM, see the USDA's publication, *Biointensive Integrated Pest Management (IPM): Fundamentals of Sustainable Agriculture*, at attra.ncat.org/attra-pub/PDF/ipm.pdf.

Table 3. The relative compatibility of some types of pesticides with biological control.

Pesticide	Compatibility with biological control
bifenazate & hexythiazox (miticides)	highly compatible
Bacillus thuringiensis (Bt)	highly compatible
carbaryl (carbamate)	not compatible
insect growth regulators	somewhat compatible
insect-parasitic nematodes	highly compatible
neem oil	compatible
neonicotinoids	somewhat compatible
organophosphates (most)	not compatible
pyrethrins (botanical)	somewhat compatible
pyrethroids (synthetic)	not compatible
soaps, oils, dusts	compatible
spinosyns	compatible

Adapted from the North Carolina State University Biological Control Information Center, cipm.ncsu.edu/ent/biocontrol/pestacides.htm.

The natural enemies of insect pests

Successful biological control programs are based on not just accurate identification of the pests and natural enemies involved, but also on a thorough understanding of their life cycles and other biological characteristics. Some natural enemies, such as adult lady beetles, are familiar to most people, but they may not be recognized in the egg, larval, or pupal stages. Other natural enemies, including most parasitic insects, are much less familiar to most people. In addition, many immature insects not only look very different from the adults of the same species, they often live in different locations, consume different foods, and require different resources or environmental conditions to prosper. Detailed knowledge of the species involved can inform your management of the various habitats inhabited by pests and natural enemies, including crop plants, soil, and noncrop vegetation.

This chapter describes the appearance, life cycles, biology, and behavior of the major groups of insects that are predators or parasites of pest insects. Spiders and predatory mites are also covered, as are insect-parasitic nematodes and insect pathogens. Birds, bats, and toads are also discussed briefly, as they can be important natural enemies, but are seldom managed for biological control. The use of these natural enemies in biological control is discussed in chapters 7–9.

Given the large number of crops grown in this region and the considerable overlap in important natural enemies among crop types, we decided to organize this chapter by type of natural enemy. We encourage you to read straight through this chapter to gain a sense of natural enemy diversity.

However, if you find it overwhelming to do so, you may want to focus on the major natural enemies of a particular crop or pest; you can find this information in Appendix 1.

Predatory and parasitic insects and arachnids

Of the 600 families of insects in the United States and Canada, about 220 include one or more species that feed on other insects, mites, spiders, millipedes, snails, slugs, nematodes, or earthworms. Many of these families, however, are of little or no consequence in pest control (figure 10). About 60–70 families contain natural enemies of pests.

Although different species of insects within a family vary somewhat in appearance and life history, they share more similarities than differences. For example, there are about 400 species of lady beetles in North America and over 2,000 species of parasitic braconid wasps. It is therefore impractical and unnecessary to produce a complete encyclopedia of every important natural enemy species. Instead, we illustrate a few examples from the more important families to show their diversity, and discuss the family characteristics in general terms. The arrangement of this section follows standard insect classification (see chapter 2), starting with the insects that have simple metamorphosis.

In the descriptions of natural enemy groups, organized summaries of essential details are followed by explanations of distinctive characteristics or specific examples for that group. The notes on conservation and augmentation will make more sense after you've read chapters 8 and 9 respectively. Species numbers represent the approximate number of species in the United States and Canada that have been described and named, and in some cases may significantly underestimate the actual number of species. These numbers (taken mainly from Arnett, R. 2000. *American Insects*, 2nd ed.) are provided to give a sense of the diversity of each group, but it's important to keep in mind that (a) not all of these species occur in a given region, and (b) many of the species in a given group may not be natural enemies of importance in agricultural settings, and some may even be plant-feeding pests.

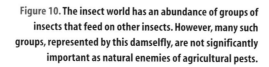

Figure 10. The insect world has an abundance of groups of insects that feed on other insects. However, many such groups, represented by this damselfly, are not significantly important as natural enemies of agricultural pests.

ORDER Mantodea: Praying mantids

FAMILY Mantidae

Species in U.S. & Canada: 18

Size: Immatures: 5–80 mm;
adults: 50–80 mm

Metamorphosis: Simple—nymphs and adults are both predatory

Common hosts/prey: Generalists—attack pests, natural enemies, and each other

Common habitats: Various

Generations per year: 1

Overwintering: A foam-like egg case, usually on plant stems, 10–30 mm in diameter, containing up to several hundred eggs (figure 12); eggs rarely survive winter in northern areas

Available for augmentation? Egg cases of Chinese mantid (*Tenodera aridifolia sinensis*) are available, but recommended only as a curiosity, not as an effective control measure

Conservation? Avoidance of broad-spectrum insecticides

Praying mantids (figure 11) are among the most recognizable of predatory insects. They have highly modified and strengthened front legs for capturing and subduing their prey, and they have chewing mouthparts. Large mantids can inflict a painful bite if handled. Many people assume that praying mantids are highly beneficial, but in fact they are opportunistic feeders: they consume whatever comes into their grasp, including other natural enemies, pollinators such as bees, innocuous insects, and each other. This renders them virtually useless as effective natural enemies of garden or crop pests.

Figure 12. Egg case of the commercially available Chinese mantid, *Tenodera aridifolia sinensis*.

Figure 11. A Chinese mantid. Praying mantids are generalist predators that are not particularly useful in pest control; their purchase and release is not recommended.

ORDER Hemiptera: True bugs

FAMILIES

- **Anthocoridae (minute pirate bugs)**
- **Miridae (plant bugs)**
- **Reduviidae (assassin bugs and ambush bugs)**
- **Lygaeidae (seed bugs and bigeyed bugs)**
- **Nabidae (damsel bugs)**
- **Pentatomidae (stink bugs)**

Although most people consider all insects and related creatures as "bugs," only insects in the order Hemiptera are technically and accurately called bugs, or true bugs. All Hemiptera undergo simple metamorphosis and have piercing-sucking mouthparts. Many true bugs are serious crop pests and others (such as bed bugs) are pests of human health or livestock. Some larger predatory Hemiptera can inflict painful bites if handled, sometimes resulting in severe inflammation of the area surrounding the bite.

In all predatory Hemiptera, both the nymphs and the adults are predatory, and they can often be found in the same general habitat feeding on similar types of prey, although the young nymphs usually require smaller prey. Some predatory true bugs also feed on plant sap, but this minor feeding does not cause plant damage. As with most generalists, predatory bugs occasionally feed on other beneficial insects, but their overall effect is usually considered beneficial.

In general, predatory Hemiptera overwinter in sheltered locations such as at the base of grasses and forages, in leaf litter, and under tree bark.

FAMILY | Anthocoridae:
Minute pirate bugs

Species in U.S. & Canada: 90

Size: Nymphs: 0.5–2 mm; adults: 1–2 mm

Metamorphosis: Simple—nymphs and adults are both predatory

Common hosts/prey: Generalists—attack aphids, thrips, psyllids, small caterpillars, leafhopper nymphs, mites, insect eggs, and more

Common habitats: Crops (especially alfalfa and corn), pastures, field margins, and habitats with herbaceous and shrubby flowering plants; several species common in stored grain bins, where they prey on stored grain pests

Generations per year: 2–3 (for most *Orius* species and *Anthocoris* species)

Overwintering stage: Adults

Available for augmentation? Several *Orius* species for aphids, thrips, and whiteflies in greenhouses; *Xylocoris flavipes* for stored grain pests; *Anthocoris nemoralis* for pear psyllids

Conservation? Reduced use of foliar and systemic insecticides; crop diversification and insectary plantings that provide plant foods and alternate prey

Most of these insects are only 1–2 mm in size, which is why they are called minute. *Orius insidiosus* (figures 13–14) is found in gardens and many crops and is probably the most important minute pirate bug in our region. *Orius tristicolor* is a similar species that is common in western states. *Orius* can destroy over 50% of the eggs of corn earworm, and both the nymphs and adults can consume 30 or more spider mites per day.

FAMILY | Miridae:
Plant bugs

Species in U.S. & Canada: 1,930

Size: Nymphs: 1–3 mm; adults 3–4 mm

Metamorphosis: Simple—nymphs and adults of beneficial species are both predatory

Common hosts/prey: Most of the 64 *Deraeocoris* species are generalist predators on spider mites, aphids, scale insects, thrips, whiteflies, lace bugs, caterpillars, and other small insects; the family includes many other predators and many plant-feeders, including some serious pests

Common habitats: *D. nebulosus* is common in orchard, landscape, and forest trees; varies for other species

Generations per year: 1 (for most species); *D. brevis* develops from egg to adult in about 30 days and adults live for about 20 days

Overwintering stage: Eggs in some species, adults in others

Available for augmentation? *D. brevis* for various pests, primarily in greenhouses; will enter hibernation when daylength is less than 10 hours or the temperature is below 73°F

Conservation? Reduced use of pesticides; presumably also by crop diversification and insectary plantings that provide alternate prey, but this has not been tested

This is a diverse family (figures 15–16) whose species have various habits. Many mirids, such as tarnished plant bug, are serious agricultural pests, but many others are predatory. It can be very difficult to distinguish predatory from pest species in this family.

Figure 15. A predatory mirid bug, *Deraeocoris* species (about 3–4 mm long).

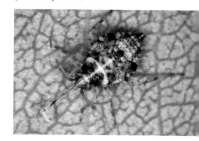

Figure 16. The nymph stage of a predatory mirid, *Deraeocoris brevis*.

Figure 13. The minute pirate bug, *Orius insidiosus*, a generalist predator of small insects and insect eggs, is often found in the garden and in a variety of crops.

Figure 14. A minute pirate bug nymph.

FAMILY | Reduviidae: Assassin bugs and ambush bugs

Species in U.S. & Canada: 160

Size: Nymphs: 2–20 mm; adults: 10–25 mm

Metamorphosis: Simple—nymphs and adults of beneficial species are both predatory

Common hosts/prey: Most are generalists that attack aphids, leafhoppers, caterpillars, beetle eggs and larvae, and other small/medium-sized insects

Common habitats: Crops, pastures, field margins, and habitats with flowering plants

Generations per year: 1

Overwintering stage: Nymphs in some species, adults in other species

Available for augmentation? No

Conservation? Reduced use of pesticides; presumably also by crop diversification and insectary plantings that provide alternate prey

Assassin bugs (figure 17) have not been specifically manipulated for biological control, but they comprise part of the generalist predator community in many settings. The ambush bugs often seen on goldenrods and other late-summer flowers are also part of this family; these do not contribute significantly to biological control, as they prey mostly on larger bees, wasps, and flies. Because of their large size, assassin bugs and ambush bugs can inflict a painful bite if handled, resulting in an inflammation that can persist for a few days.

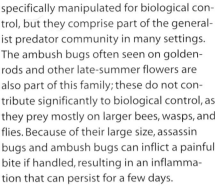

Figure 17. An assassin bug, *Acolla multispinosa*.

FAMILY | Lygaeidae: Seed bugs and bigeyed bugs

Species in U.S. & Canada: 320

Size: Nymphs: 1–3 mm; adults: 3–4 mm

Metamorphosis: Simple—both nymphs and adults of beneficial species are predatory

Common hosts/prey: *Geocoris* species are generalists that attack spider mites, aphids, scale insects, whiteflies, lace bugs, caterpillars, chinch bugs, and other small insects; most other lygaeids feed on seeds and/or fruits and some may be pests

Common habitats: Crops, pastures, field margins, turf grass, and habitats with flowering plants

Generations per year: Probably 2–3

Overwintering stage: Eggs or adults

Available for augmentation? *Geocoris punctipes* for various pests, but primarily for aphids in greenhouses

Conservation? Reduced use of pesticides; crop diversification and insectary plantings that provide plant foods and alternate prey

The genus *Geocoris* includes 25 species in North America, including several that are common in many noncrop situations and in agricultural settings where few broad-spectrum insecticides are used. In turf grass, bigeyed bugs prey on chinch bugs, greenbugs, sod webworm, and other turf grass pests. *G. punctipes* is a small insect, about 3–4 mm long, but it can feed on as many as 1600 spider mites during the course of its nymphal development and an additional 80 mites per day as an adult (figure 18).

Figure 18. Bigeyed bug, *Geocoris* species.

FAMILY Nabidae: Damsel bugs

Species in U.S. & Canada: 34

Size: Nymphs: 1–12 mm; adults: 9–13 mm

Metamorphosis: Simple—both nymphs and adults are predatory

Common hosts/prey: Generalists—attack leafhoppers, aphids, small caterpillars and beetle larvae, moth eggs, and other small insects

Common habitats: Most common in gardens and field and row crops, especially in association with grasses, alfalfa, and soybeans

Generations per year: 1–5

Overwintering stage: Adults for most *Nabis* species, eggs for at least some other genera

Available for augmentation? No

Conservation? Reduced use of pesticides; presumably also by crop diversification and insectary plantings that provide alternate prey

Three species in the genus *Nabis* (figures 19–20) are the most abundant damsel bugs in crops in the upper Midwest. These are all tan-colored bugs and are similar in appearance. Some other species of nabids are black, but these are less common in agricultural settings.

FAMILY Pentatomidae: Stink bugs

Species in U.S. & Canada: 260

Size of adults: 10–15 mm

Metamorphosis: Simple—both nymphs and adults of beneficial species are predatory

Common hosts/prey: *Podisus maculiventris* (spined soldier bug) is a generalist, but especially attacks Colorado potato beetle, caterpillars, and plant-feeding true bugs; *Perillus bioculatus* (twospotted stink bug) is more specialized on eggs and larvae of Colorado potato beetle and other leaf beetles; most other species are plant-feeders and may be pests

Common habitats: *P. maculiventris*—field crops, meadows, shrubs, forests; *P. bioculatus*—potato fields

Generations per year: 1–3 (for *P. maculiventris* and *P. bioculatus*); varies for other species

Overwintering stage: Adults for *P. maculiventris* and *P. bioculatus*

Available for augmentation? *P. maculiventris* for Colorado potato beetle and other pests; its aggregation pheromone is also available to attract *P. maculiventris* from surrounding areas into fields

Conservation? Reduced use of pesticides; presumably by crop diversification and insectary plantings that provide alternate prey

Adult pentatomids have a broad, shield-shaped body. They are usually green or brown, but some are brightly colored. Many discharge a disagreeable odor when handled. The spined soldier bug, *Podisus maculiventris,* (figures 21–23) and the twospotted stink bug, *Perillus bioculatus,* (figure 24) are highly efficient predators capable of consuming many prey during the course of their development. Both species have been introduced into France for biological control of Colorado potato beetle. *Podisus* females lay clusters of about 30 barrel-shaped eggs on leaves, and each female can produce over 1000 eggs in her lifetime. Nymphs are small and almost circular in shape.

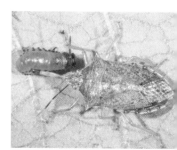

Figure 22. Adult spined soldier bug beginning to feed on Colorado potato beetle larva.

Figure 19. An adult damsel bug, *Nabis* species.

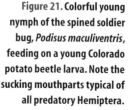

Figure 21. Colorful young nymph of the spined soldier bug, *Podisus maculiventris*, feeding on a young Colorado potato beetle larva. Note the sucking mouthparts typical of all predatory Hemiptera.

Figure 23. Adult spined soldier bug completes feeding on larva in about 20 minutes.

Figure 20. A damsel bug nymph.

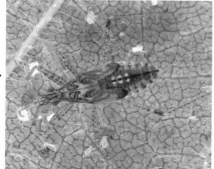

Figure 24. Adult twospotted stink bug, *Perillus bioculatus*.

ORDER ▶ Coleoptera: Beetles

FAMILIES

- **Carabidae (ground beetles)**
- **Coccinellidae (lady beetles or ladybird beetles)**
- **Staphylinidae (rove beetles)**
- **Histeridae (hister beetles)**
- **Other families of lesser importance**

The beetles are the most diverse group of insects in the world, constituting about 40% of all known insect species. There are about 27,000 known species in the United States. All beetles undergo complete metamorphosis, with egg, larval, pupal, and adult stages. The front wings are hardened structures called elytra, which protect the hind wings and abdomen when the insect is not in flight.

Beetles live in water and soil, in association with plants, and in other diverse habitats. Their foods also vary: many are scavengers of dead plant and animal matter, others feed on plants or wood, others are predatory or parasitic. Depending on the species, larvae and adults may do similar or very different things. For example, both the larvae and adults of most lady beetle species are predatory, usually on the same types of prey. However, blister beetle larvae are parasitic on grasshopper eggs and ground-nesting bees, whereas the adults are generally plant feeders.

Approximately 40 families of beetles are known to have predatory members, but some of these groups feed on insects only incidentally. By far, the two most important families in crop protection are the predatory ground beetles (family Carabidae) and the lady beetles (family Coccinellidae).

FAMILY ▷ Carabidae: Ground beetles

Species in U.S. & Canada: 2,270

Size of adults: 3–40 mm

Metamorphosis: Complete—adults and larvae of most species are both predatory

Common hosts/prey: Most species are generalists—larvae and adults prey on various insects, earthworms, and other small organisms in or on soil or in vegetation

Common habitats: Soil, especially in arable crops and heavier soils; many species climb into vegetation at night; some live in trees and shrubs

Generations per year: Usually 1; a few species require more than a year to develop; some adults can live 2–4 years

Overwintering stage: Adults or larvae in hedgerows, sod, or litter

Available for augmentation? No

Conservation? Reduced tillage; avoiding soil insecticides and fumigants; providing undisturbed sod mulch or leaf litter near crops for overwintering habitat

Adult carabid beetles (figures 25–26) are usually brown or black, but a few are metallic blue or green. Most species are predatory as both larvae and adults, although some are scavengers and a few feed on plants and seeds. Many insects, even leaf-feeding insects, pass part of their life cycle in the soil or under leaf litter, especially in the pupal stage or while overwintering, and during this stage they may be attacked by ground beetles.

Some ground beetles climb into trees, shrubs, and crop plants in search of prey. Beetles of the genus *Calosoma* are called caterpillar hunters because they are voracious predators of caterpillars. The adult beetles are among the largest in the family, and both adults and larvae are very active predators. *C. sycophanta* (figure 26) is a large, metallic green beetle introduced into the United States from Europe for control of gypsy moth. A larva (figure 27) can consume as many as 40 large gypsy moth larvae, while adults can kill 200 or more.

Figure 26. A caterpillar hunter, *Calosoma sycophanta*. These large ground beetles (20–35 mm) were introduced into the United States from Europe for control of gypsy moth.

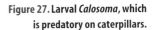

Figure 27. Larval *Calosoma*, which is predatory on caterpillars.

Figure 25. The ground beetle *Harpallelus basalaris* (family Carabidae). Members of this common family feed on insects found in or on the soil.

FAMILY Coccinellidae: Lady beetles or ladybird beetles

Species in U.S. & Canada: 400

Size of adults: 2–8 mm

Metamorphosis: Complete—adults and larvae of most species are both predatory

Common hosts/prey: Most species are generalists on aphids, scale insects, mealybugs, and whiteflies, sometimes spider mites and insect eggs; some attack small caterpillars and leaf beetle larvae; adults may also feed on nectar and pollen; *Cryptolaemus montrouzieri* specializes on mealybugs; *Stethorus* species specialize on spider mites and other mites; a few (e.g., Mexican bean beetle) feed on plants or fungi

Common habitats: Associated with plants—preferences for types of herbaceous or woody plants vary by species

Generations per year: 2–3

Overwintering stage: Adults, in most species

Available for augmentation? *Hippodamia convergens* (though not recommended) and *Coccinella septempunctata* for aphids and other pests; *Stethorus punctillum* for spider mites; *Cryptolaemus montrouzieri* (mealybug destroyer) for mealybugs; *Delphastus* species for whiteflies; *Pseudoscymnus tsugae* for wooly hemlock adelgid

Conservation? Reduced use of pesticides; crop diversification and insectary plantings that provide pollen and alternate prey

The lady beetles are a diverse group of important natural enemies. Although frequently called ladybugs, these insects are not true bugs and therefore the other common names are preferred. The first successful case in modern biological control involved a lady beetle, the vedalia beetle (see page 2).

The common perception is that lady beetles are somewhat oval in shape, and reddish colored with black spots. While the oval shape is consistent within this family of beetles, coloration varies greatly, from pink to red to orange to black. A close look at the beetles reveals that the number and pattern of spots varies; this often relates to the different species. Most species tend to specialize on aphids as prey, but they will eat other soft-bodied insects as well. Some lady beetles are pure black, or even black with red spots. Many of these beetles specialize in other types of prey as noted in the next few paragraphs.

Lady beetles overwinter as adults. Some species, such as our native convergent lady beetle, *Hippodamia convergens*, congregate in enormous hibernation clusters, especially in the western states. Other species overwinter singly or in small clusters. In spring they seek out the aphids or other prey that will be food for both adults and larvae. Eggs are laid adjacent to the prey. Many species deposit oblong eggs on the leaf surface. Some species scatter individual eggs while other species lay compact clusters of 10–20 eggs (figure 28). Eggs of the aphid-feeding species are usually yellow to orange in color, and 1–1.5 mm long.

Lady beetle larvae (figures 29–30) are also important predators, but they are not as familiar as the adults to many people. The larvae of the aphid-feeding species are somewhat slender, with the body tapering to a blunt point at the rear. Depending on species and instar, they are 3–15 mm long. The color is usually black or dark gray, often with conspicuous, red, yellow, orange, or blue markings. The prominent legs are held out to the sides. Lady beetle larvae that feed on mealybugs and scale insects may not be as conspicuous, as they may be covered in a white waxy coating similar to that of the prey insects. A larval lady beetle normally consumes 500–1000 aphids or similar prey during its growth.

Figure 30. Larva of the multicolored Asian lady beetle, *Harmonia axyridis*.

Figure 28. Twospotted lady beetle laying eggs.

Figure 29. Young larvae of the lady beetle *Coleomegilla maculata*.

The entire life cycle takes about 4–6 weeks. Eggs usually hatch in 3–7 days. If prey are abundant and temperatures warm, most lady beetle larvae complete development in 2–4 weeks. They pupate where they were feeding (figure 31). The pupal stage, which lasts about a week, also is unfamiliar to most people.

Many species of beneficial lady beetles occur in the North Central United States. The convergent lady beetle (figure 32), is one of the most common; it is an important predator of aphids and other pests. The seven-spotted lady beetle, *Coccinella septempunctata*, (figure 33), a relatively large lady beetle from Europe, has been distributed in the United States for aphid control. It has rapidly spread throughout much of the North Central region.

Harmonia axyridis, the multicolored Asian lady beetle, (figure 34) was released in the United States multiple times between 1916 and 1985. Because it was not seen again in this country until 1988, when it was collected in Louisiana and Mississippi, some suspect that the successful establishment of *H. axyridis* was actually due to an accidental introduction, perhaps as a stowaway on a cargo ship. In any case, *H. axyridis* has now spread throughout much of the United States and southern Canada, becoming one of the most important predators of aphids and several other pests in a variety of crops. However, *H. axyridis* has also caused several adverse effects. It has reduced populations of other lady beetle species, and has become a familiar nuisance pest due to its overwintering aggregations in buildings. In addition, it has been blamed for damage to fall-ripening fruit, though apparently, with the exception of raspberries, it can only feed on fruit with existing injuries. Adult beetles are difficult to remove from clusters of grapes, and even small numbers of beetles can result in tainted wine. It was commercially available for a few years, but has been taken off the market due to some of these problems.

Figure 31. Pupa of the variegated lady beetle, *Hippodamia variegata*. Some lady beetle pupae are darker in color.

Figure 32. The convergent lady beetle, *Hippodamia convergens*.

Figure 34. The multicolored Asian lady beetle, *Harmonia axyridis*, gets its name because the color can range from black to orange to red, and the spotting pattern can be variable, or even absent.

Figure 33. The seven-spotted lady beetle, *Coccinella septempunctata*, is an introduced species.

Except for perhaps *H. axyridis*, *Coleomegilla maculata* (figure 35) is often the most common lady beetle in corn, potatoes, and mixed vegetable crops, where it feeds on mites, aphids, caterpillar eggs, and eggs and small larvae of asparagus beetles, Colorado potato beetles, and Mexican bean beetles. Pollen may comprise up to half of its diet, so it is particularly responsive to insectary plantings and other conservation practices that provide pollen sources.

The twicestabbed lady beetle, *Chilocorus stigma*, is small and black, with a bright red spot on either side of its body. It is an important predator of scale insects and other pests. It is frequently seen in association with cottony maple scale and undoubtedly is important in control of this pest of silver maples. *Chilocorus kuwanae* is similar in size and coloration and has been introduced into the United States for control of euonymus scale. *Pseudoscymnus tsugae* is available for use against the introduced wooly hemlock adelgid.

Members of the genus *Stethorus* (figure 36), often called spider mite destroyers, are only a few millimeters long and black in color, and therefore are not very conspicuous. Larvae consume 200 or more spider mites during their development, and adults can eat up to 100 mites per day. One species is common on apple foliage, and at least one species is commercially available for augmentation.

The mealybug destroyer, *Cryptolaemus montrouzieri* (figures 37–38), was introduced into the United States for controlling mealybugs on citrus and other crops. Although it does not survive winters in most of the United States, it is commercially produced and can be purchased for periodic release. In the North Central United States it has been most useful for mealybug control in greenhouses, conservatories, and interior plantscapes.

Several other lady beetle species are commercially available (see "Available for augmentation?" on page 31); most of these can be used effectively in augmentative releases, especially in greenhouses and conservatories, but the convergent lady beetle, *H. convergens*, is generally not recommended because it disperses rapidly following release (this behavior is discussed in more detail in chapter 9).

Figure 36. Spider mite destroyer, *Stethorus punctum*, adult (top) and larva (bottom). This tiny black lady beetle feeds primarily on spider mites.

Figure 35. *Coleomegilla maculata* is a common lady beetle in the Midwest.

Figure 37. The mealybug destroyer, *Cryptolaemus montrouzieri*, is a specialist lady beetle.

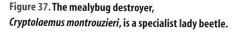

Figure 38. The mealybug destroyer larva has a waxy covering that makes it resemble its prey.

FAMILY ▸ Staphylinidae: Rove beetles

Species in U.S. & Canada: 3,200

Size of adults: 3–30 mm

Metamorphosis: Complete—both larvae and adults of most species are predatory

Common hosts/prey: Generalists—mites, insect eggs and small larvae

Common habitats: Most soils, mulch, compost, and other decaying organic material; some species found on vegetation

Generations per year: Varies

Overwintering stage: Varies

Available for augmentation? No

Conservation? Reduced tillage; avoiding soil insecticides and fumigants; providing undisturbed sod mulch, or leaf litter near crops for overwintering habitat

This is the most diverse family of North American beetles, but is not well studied. Most are thought to be predatory, some have larvae that are parasites, and many others are probably scavengers. These insects are usually less than 6 mm long, and they tend to hide in soil or debris. Even so, they are quite recognizable (figure 39) because of their slender, usually black body, shortened front wings (elytra), and behavior of curling the tip of the abdomen upwards when disturbed or running.

Figure 39. A rove beetle, family Staphylinidae.

FAMILY ▸ Histeridae: Hister beetles

Species in U.S. & Canada: 500

Size of adults: < 10 mm

Metamorphosis: Complete—both larvae and adults of most species are predatory

Common hosts/prey: Fly eggs and larvae for dung-dwelling species, bark beetles and wood borers for bark-dwelling species

Common habitats: Dung, carrion, compost, beneath tree bark

Generations per year: Varies

Overwintering stage: Varies

Available for augmentation? *Carcinops pumilio* for filth fly eggs and larvae, especially in poultry houses

Conservation? Natural enemy suppliers can describe manure management practices that conserve hister beetles and other natural enemies.

Hister beetles are usually less than 10 mm long, shiny black, and often somewhat broad and flattened. Members of the genus *Platysoma* (figure 40) live under tree bark where they are important predators of bark beetles. Larvae of *Carcinops pumilio* (figure 41) consume 2–3 house fly eggs per day, and adults can consume 13–24 eggs per day. At 77°F, the complete life cycle takes about 40 days.

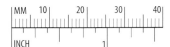

Figure 40. A hister beetle, *Platysoma* (=*Cylistix*) *cylindrica*, a predator of bark beetles. Many hister beetles are broadly oval in shape but this species is more cylindrical, allowing it to crawl through bark beetle tunnels.

Figure 41. *Carcinops pumilio* (family Histeridae), in both adult and larval stages, is a predator of the eggs and larvae of filth flies such as the house fly.

Additional families of beetles

A few additional families have at least some species that are natural enemies of pests, but these are generally less common and less important than the beetles described above. These include:

Cantharidae

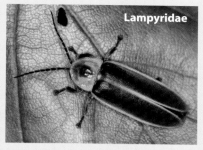
Lampyridae

Cantharidae = soldier beetles (470 species): Adults feed on pollen, nectar, and soft-bodied insects; larvae prey on insect eggs and larvae, snails, slugs, millipedes and earthworms in organic litter or under bark.

Lampyridae = fireflies (125 species): Larvae prey mostly on snails and slugs, but may also attack earthworms, cutworms, and various other soft-bodied insects in soil and other moist habitats; some adults are also predators.

Cleridae

Lycidae

Cleridae = checkered beetles (270 species): Larvae of most species prey on larvae of bark beetles, longhorned beetles, and similar pests of trees.

Lycidae = netwinged beetles (80 species): Larvae prey on other insect larvae and pupae, mostly in soil and under bark.

Elateridae

Meloidae

Elateridae = click beetles (890 species): Larvae of some species are generalist predators under bark or in soil or rotting logs.

Meloidae = blister beetles (310 species): Larvae of many species feed on grasshopper eggs, others on ground-nesting bees; adults of some species may be plant pests and/or toxic to livestock.

Nitidulidae = sap beetles or picnic beetles (180 species): A few species prey on bark beetles; some are attracted to European corn borer tunnels and have been tested as delivery agents for the insect-pathogenic fungus, *Beauveria bassiana*. The scale picnic beetle, *Cybocephalus nipponicus*, is commercially available for controlling scale insects.

ORDER Neuroptera: Lacewings, antlions, and others

FAMILIES

- **Chrysopidae (green lacewings)**
- **Hemerobiidae (brown lacewings)**

The order Neuroptera contains several small families, most of which are predatory or parasitic as larvae and predatory as adults. Several families are quite uncommon, and others, such as the antlions, owlflies, and snakeflies are more common in the South and West than in northern states. Two families of lacewings provide appreciable benefit to agriculture in the North Central region.

FAMILY Chrysopidae: Green or common lacewings

Species in U.S. & Canada: 90

Size of adults: 12–20 mm

Metamorphosis: Complete—both larvae and most adults are predatory

Common hosts/prey: Larvae feed mostly on aphids, but also on mites, thrips, mealybugs, caterpillars, and other small soft bodied insects and their eggs; adults of some species are predatory, others feed only on pollen and honeydew

Common habitats: Different species are found in open fields, brushy habitats, and trees

Generations per year: 1–3

Overwintering stage: Some species as adult, some as mature larva in cocoon

Available for augmentation? *Chysoperla carnea, C. rufilabris,* and perhaps other species are available as larvae or as eggs in a bran carrier (sprinkle on crop), mostly for aphid control in protected areas and greenhouses

Conservation? Reduced use of pesticides or use of "softer" chemicals; crop diversification and insectary plantings that provide plant foods and alternate prey; food sprays can attract adults and induce egg-laying

Green lacewing adults (figure 42) are most active at night, but are also commonly seen during daylight. They are easily recognized by their soft, slender, bright green bodies, golden eyes, large clear membranous wings, and long hair-like antennae. In North America, the genus *Chrysoperla* has the most species; *Chrysopa* is another genus with important natural enemies, but these two genus names are sometimes used interchangeably, so don't let that confuse you. Commercially available species differ in their environmental requirements and prey preferences, so confirm your choice of species with your biological control supplier, should you choose to purchase green lacewings.

Figure 42. An adult green lacewing, *Chrysoperla* species.

Figure 43. The distinctive stalked eggs of a green lacewing. The stalk is about 6–8 mm high. Some species lay their eggs in clusters, others lay them singly.

Figure 44. Newly hatched larvae of green lacewings, called aphidlions.

Green lacewing eggs are also easily identified, as an oval, white egg attached to an upright hair-like stalk about 8 mm long (figure 43). The eggs are usually laid on foliage near colonies of aphids or other prey. Most species lay their eggs singly, but some lay them in clusters. Eggs hatch into small, gray, slender larvae called aphidlions, and can consume up to 400 aphids per week (figures 44–45). These larvae have enlarged, sickle-shaped mouthparts that extend forward from the front of the head; these are used to puncture the prey and suck out the internal fluids. The larva ultimately grows to about 10 mm long, then spins a spherical, silken cocoon within which it pupates, usually on the underside of a leaf (figures 46–47).

Figure 45. A grown green lacewing larva.

Figure 46. The silken cocoon of green lacewings is usually spherical.

Figure 47. The pupa of a green lacewing, normally encased in a silken cocoon.

FAMILY   Hemerobiidae: Brown lacewings

There are about 60 species of brown lacewings in North America north of Mexico. They are similar to green lacewings in general appearance, but adults are smaller (6–12 mm) and brown in color (figure 48), and the eggs (figure 49) are not stalked. Brown lacewing larvae (figure 50) are more slender and delicate-looking than green lacewing larvae. Brown lacewings occur in fields and forests, and are less common than green lacewings in crops and gardens. Both larvae and adults feed on aphids and other small, soft-bodied insects. None are available commercially in the United States.

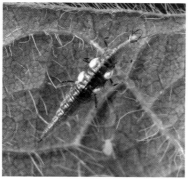

Figure 48. A brown lacewing, family Hemerobiidae.

Figure 49. A brown lacewing egg.

Figure 50. A grown larva of a brown lacewing.

ORDER | Diptera:
Flies, gnats, midges, and others

FAMILIES

- **Cecidomyiidae (gall midges)**
- **Syrphidae (hover flies)**
- **Tachinidae (tachinid flies)**
- **Other families of lesser importance**

The flies are a highly diverse group of insects; different species show a wide array of habits and occupy many different habitats. Approximately 35 families contain species that are predatory or parasitic on other insects. Some of these occur primarily in aquatic or semiaquatic environments where they feed on such insects as mosquitoes, black flies, and other public-health or nuisance pests. Only families that are common or important in agriculture are discussed here.

FAMILY | Cecidomyiidae:
Gall midges

Species in U.S. & Canada: 1,060 (most are gall-making plant-feeders; discussion here pertains only to beneficial species)

Size of adults: tiny, < 4 mm

Metamorphosis: Complete—larvae of beneficial species are predatory

Common hosts/prey: Mostly aphids, but also scale insects and mites

Common habitats: Larvae are found amidst aphid colonies or spider mite outbreaks on plants

Generations per year: 3–6 in the field, more in a greenhouse, especially with nighttime lighting

Overwintering stage: Pupa in soil

Available for augmentation? *Aphidoletes aphidimyza* for aphids, primarily in greenhouses; *Feltiella acarisuga* for spider mites; both are shipped as pupae within cocoons in a moist carrier designed for sprinkling on damp soil

Conservation? Provide damp soil for pupation and humid air (>80% relative humidity); in greenhouses, supplemental lighting at night will prevent hibernation during shorter days of fall and winter; presence of non-pest aphids can boost numbers of predatory gall midges before pest aphids develop

A majority of the species in this family feed on plants, inducing abnormal plant growths called galls. However, the larvae of some species feed on small insects and mites in galls, in decaying organic matter, or on plant surfaces. The gall midges are all tiny, slender-legged, rather delicate insects, and the larvae are legless maggots. Adults and fully grown larvae are rarely over 3–4 mm long, and the eggs may be as small as 0.1 mm ($^1/_{250}$ inch!). Because they are so small, these predators are frequently overlooked.

Aphidoletes aphidimyza is commercially available for aphid control, primarily in greenhouses. A larva feeds by injecting a toxin into an aphid to paralyze it, then sucking out the body fluids of the immobilized aphid (figure 51). Full-grown larvae are bright orange, but very small (<3 mm). They often feed on aphids much larger than themselves and may kill more aphids than they actually consume. The exoskeletons of aphids killed by *A. aphidimyza* often remain attached to the host plant, which could be an aesthetic problem in some situations. Under suitable conditions in a greenhouse, just one or a few releases of *A. aphidimyza* at weekly intervals can lead to season-long aphid control. *Aphidoletes* also is commonly seen feeding on aphids in agricultural settings, such as in soybean fields feeding on soybean aphids.

Aphidoletes thompsoni was successfully introduced from Europe into Canada and the northern United States for control of balsam woolly adelgid, but it seems to have had little effect on populations of the pest.

Figure 51. The larva of a commercially available predatory midge, *Aphidoletes aphidimyza*, feeding on a green peach aphid.

Figure 52. Adult hover fly, genus *Toxomeris*.

FAMILIES

Syrphidae: Hover flies

Species in U.S. & Canada: 870

Size of adults: 6–18 mm

Metamorphosis: Complete—larvae of beneficial species are predators

Common hosts/prey: Larvae feed on aphids as well as small caterpillars, thrips, scales, and other small insects; adults feed primarily on nectar and pollen

Common habitats: Adults are often seen at flowers. Larvae live in diverse habitats, including decaying vegetation and stagnant or polluted water; many predatory species occur on vegetation in open fields, crops, and gardens

Generations per year: 4–7 (typical life cycle takes 2–4 weeks)

Overwintering stage: Pupae in soil or plant debris

Available for augmentation? No

Conservation? Provide appropriate flowering plants within or adjacent to crops to provide pollen and nectar for adults (females require pollen for egg production); some species avoid windy areas, so windbreaks may be useful

Some species of adult hover flies (figures 52–53) are very common and easily identified by their hovering flight and generally bee-like appearance. Although they may be striped yellow and black like a small bee or wasp, they cannot bite or sting.

Although most adult syrphid flies visit flowers for nectar and pollen, the habits of the larvae are very diverse. Many, however, are predators of aphids and other insects that occur on plants. The females of the aphid-feeding species deposit small white eggs on the plant surface amidst growing aphid colonies. Pale green to yellow larvae emerge from the eggs. Larvae are maggots, with poorly developed fleshy legs and lacking distinct heads (figures 54–55). Each larva consumes as many as 400 aphids during its development, and can grow to 10–15 mm long in the larger species. Casual examination can underestimate syrphid predation, since the larvae often move to the base of the plant or another sheltered location during daylight hours, then return to the aphid colony at night to resume feeding. Similarly, some species pupate on the plant surface near the feeding site, whereas others enter the soil to pupate. Either way, the pupa is enclosed within a smooth, green or tan, teardrop-shaped **puparium** (figure 56), the hardened skin of the last larval instar.

Figure 53. Adult hover fly, genus *Syrphus*.

Figure 54. Hover fly larva feeding on an aphid.

Figure 55. Larva of a different species of hover fly.

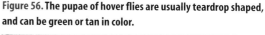

Figure 56. The pupae of hover flies are usually teardrop shaped, and can be green or tan in color.

Figure 57. This tachinid fly parasitizes tent caterpillars.

Figure 58. Tachinid fly eggs on the front of a tent caterpillar. When the eggs hatch, the parasite larvae immediately enter the caterpillar and begin feeding within. Not all tachinid species lay their eggs on the surface of a host; see text for other behaviors.

Figure 59. Fully grown tachinid larva emerging from the tent caterpillar host, which it has killed.

Figure 60. Puparium of the tent caterpillar parasite, next to its dead host caterpillar.

FAMILY | **Tachinidae: Tachinid flies**

Species in U.S. & Canada: 1,280

Size of adults: 5–15 mm

Metamorphosis: Complete—larvae are parasites

Common hosts/prey: Most species are fairly host-specific—caterpillars and beetle larvae and adults are the most common hosts, but sawfly larvae, true bugs, grasshoppers, and other hosts are also attacked

Common habitats: Wherever hosts are found

Generations per year: 1–3 or more

Overwintering stage: Varies by species

Available for augmentation? No

Conservation? Reduced use of pesticides; crop diversification and insectary plantings that provide plant foods and alternate hosts

The Tachinidae is by far the most diverse and important family of flies that parasitize insects. Larvae of all species are parasitic, and many are important natural enemies of major pests. Many foreign species of tachinids have been introduced into North America to suppress populations of non-native pests. Tachinid flies differ in color, size, and shape (figures 57, 61–62), but many resemble house flies. They usually are gray, black, or striped, and often are quite hairy or bristly, but others have red markings and some have few noticeable hairs.

The life cycles of many tachinids have been studied in detail. A species may use one of four methods to attack its hosts. Some species deposit small eggs on foliage and the eggs do not hatch until they are ingested by a host, such as a caterpillar. Others place eggs containing well-developed embryonic larvae onto foliage; the eggs hatch shortly after being deposited and the newly emerged maggots either actively search for a host or sit in ambush awaiting a host. In other species, the adult fly glues her eggs to the body of the host (figure 58). After the eggs hatch, the maggots penetrate the host's body. Some adult female tachinids possess a piercing ovipositor that allows them to insert eggs directly into the host's body. In all cases, the larva feeds inside the host's body, emerging from the dead host when its development is complete (figure 59).

Figure 61. The tachinid fly, *Arcytas apicifer*, is a parasite of cutworms and armyworms.

Egg and larval development are rapid for most tachinids, and pupation often occurs within 4–14 days after egg laying. The pupal stage (figure 60), which lasts about 1–2 weeks, usually is found as a puparium in soil or, less often, within or adjacent to the remains of the dead host. Many species are capable of several generations per year, but others are restricted to only one generation, especially if their hosts have only a single generation. Most species are **solitary parasites**, with just one parasite larva per host, but others are **gregarious parasites**, with 2–12 maggots developing within a single host.

Figure 62. *Gymnosoma* species (family Tachinidae) parasitize nymphal and adult stink bugs.

Additional families of flies

The following families are occasionally important in natural and biological control, but they are generally less common and less important than the flies already described.

Asilidae

Asilidae = robber flies (880 species): Adults are aggressive generalist predators that capture large prey while in flight; larvae prey on soft-bodied insects, larvae, and eggs in soil or decaying wood. Most common in the southwestern United States.

Bombyliidae

Bombyliidae = bee flies (800 species): Larvae parasitize cutworms, beetle grubs, bee and wasp larvae, or grasshopper egg pods. Most common in the southwestern United States.

Chamaemyiidae = aphid flies (36 species): Larvae of some species are predators of aphids and soft scales.

Dolichopodidae

Dolichopodidae = long-legged flies (1,230 species): Adults are active predators of small soft-bodied insects, including aphids; larvae are predators in decaying logs, moist organic soil, or other moist habitats.

Phoridae = hump-backed flies (225 species): Larvae of many species parasitize spiders, millipedes, wasps, bees, beetles, caterpillars, crickets, termites, ants, snails and slugs. Some show promise in classical biological control of the imported fire ant.

Pipunculidae

Pipunculidae = big headed flies (100 species): Larvae parasitize leafhoppers and planthoppers.

Pyrgotidae

Pyrgotidae = pyrgotid flies (5 species): Larvae parasitize white grubs in soil.

Sarcophagidae = flesh flies (330 species): Many feed as larvae in decaying animal flesh, but the larvae of many species are parasitoids of beetles, grasshoppers, caterpillars, or other insects; *Sarcophaga aldrichi* is an important parasite of forest tent caterpillar.

ORDER

Hymenoptera:
Wasps, ants, and bees

FAMILIES

- **Braconidae (braconid wasps)**
- **Ichneumonidae (ichneumons or ichneumonid wasps)**
- **Chalcidoidea (chalcid wasps and their relatives)—SUPERFAMILY**
- **Vespidae and Sphecidae (stinging wasps)**
- **Formicidae (ants)**
- **Other families of lesser importance**

The Hymenoptera is the second-most diverse order of insects (after the beetles) and is the most beneficial. Almost 18,000 species, most of which are beneficial, are known from the United States and Canada. Honey bees provide honey and wax and pollinate our crops, but many other species are important pollinators. Ants and some of the larger stinging wasps, such as yellowjackets and hornets, are important predators of caterpillars and other pests. The parasitic wasps are the most important group of natural enemies of pest insects, by far, with an estimated 65,000 species worldwide. Many species have been transported around the world for classical biological control of non-native pests, and numerous species are commercially available for augmentative biological control. Many species of parasitic wasps are tiny and easily overlooked. Because of their small size, many

are still unknown to science, and new species are discovered each year.

All Hymenoptera undergo complete metamorphosis, having egg, larval, pupal, and adult stages. In parasitic wasps, the larval stage develops in and kills a single host insect. Most adult parasitic wasps have a very high reproductive capacity; individual females can produce many hundreds of offspring, each of which will kill one host insect. Because they reproduce so rapidly, parasitic wasps efficiently overtake increasing host populations and often are able to suppress pests before they reach injurious levels. Some adult parasitic wasps also feed on insects, usually the same species that are host to the larval stage; this behavior is called **host feeding**.

Many adult wasps also require sugars and proteins as nutritional sources. Adults of these species feed on pollen, nectar, and honeydew, which is a sweet fluid excreted by sap-feeding insects such as aphids and scale insects. Providing pollen- and nectar-producing plants can help these species thrive (see chapter 8).

The behavior of parasitic Hymenoptera is quite complex. Most species attack only one host species or a narrow range of similar hosts. The adult females are very efficient at locating the specific hosts that will be the food for their offspring. Adult female wasps are able to locate and parasitize hosts at low host densities, which is important for keeping pest populations below economically damaging levels.

As a group, parasitic Hymenoptera use almost all groups of terrestrial insects as hosts, though most species have a limited host range of one or a few closely related species. Insect groups important to agriculture, such as caterpillars, beetles, sawflies, aphids, and scale insects, are frequent hosts. Depending on the parasite species, virtually any host stage can be attacked (egg, nymph, larva, pupa, or adult).

Most parasitic wasps cannot sting humans, unlike their larger, more familiar relatives, such as yellowjackets and bald-faced hornets. These larger wasps, easily recognized because of their size, bright colors, and ability to sting, are predators of insects.

Figure 63. *Bracon hebetor* is a braconid wasp that parasitizes moth larvae in stored grain. Here several *Bracon* larvae are shown feeding on a dead Indian meal moth larva.

Figure 64. An adult *Cotesia glomerata* about to parasitize an imported cabbageworm.

Figure 66. *Macrocentrus grandii*, a braconid parasite of European corn borer larvae.

Figure 65. Cocoons of the braconid, *Cotesia glomerata,* attached to a dead imported cabbageworm larva.

<table>
<tr><td>FAMILY</td><td></td></tr>
</table>

FAMILY **Braconidae:
Braconid wasps**

Species in U.S. & Canada: 2,050 (including subfamily Aphidiinae)

Size of adults: 2–12 mm

Metamorphosis: Complete—larvae are parasitic

Common hosts/prey: Each species is fairly host-specific—caterpillars are most common hosts, but flies, sawflies, wood-boring beetles, weevils, leafmining insects, true bugs, and ants are also attacked; members of the subfamily Aphidiinae (figures 68–69) are important parasites of aphids

Common habitats: Same habitats as their hosts; adults of many species use honeydew or flower nectar as a food source

Generations per year: Varies; 1 to several

Overwintering stage: Usually as egg or larva within host or as pupa within cocoon

Available for augmentation? A few of the many species that are available include: *Bracon hebetor* for stored product pests; *Aphidius* species, *Diaeretiella rapae,* and *Lysiphlebus testaceipes* for aphids; *Dacnusa sibirica* for agromyzid leafminers; *Cotesia* species for certain caterpillars

Conservation? Reduced use of pesticides; crop diversification and insectary plantings that provide plant foods and alternate hosts

All known species of this diverse family (figures 63–68) are parasites. Braconids usually parasitize the larval (or nymphal) stages of their hosts. However, females of a few species lay eggs in the eggs of the host insect and the parasite eggs do not hatch until the host is in the larval stage. A few braconids parasitize the adult stage of their hosts. For example, *Microctonus aethiopoides* parasitizes and sterilizes adult alfalfa weevils. Most braconids are parasitic inside the bodies of their hosts, but some attach to the outside surface of the host's body and feed through its exoskeleton (figure 63).

The number of parasite generations per year does not necessarily relate to the number of host generations. Some braconids can complete two or more generations during a single generation of the host. Others have only one generation per year. The braconid life cycle is relatively short, with most developing from egg to adult in 10–30 days. Almost all braconid larvae leave the host just prior to pupation, which usually occurs within a silken cocoon (figure 65).

Some braconids are solitary parasites, whereas others are gregarious, with several parasites developing from a single host insect (figures 63 and 65). Solitary species tend to be about the same size as their hosts, but the gregarious species are much smaller than their hosts. Most are black, brown, or tan; some have red, yellow, or orange markings. The females of many species have a noticeable elongate, needle-like, ovipositor used for laying eggs; although it may look like a stinger, it cannot be used to sting humans.

Braconids are very important in the natural or biological control of serious agricultural and forestry pests. Several non-native species of braconids have been successfully introduced into North America for the control of introduced pests.

Table 4. Some species of braconid wasps important in the North Central United States.

Parasite	Pest(s) attacked
Apanteles melanoscelus	gypsy moth
Bracon hebetor	moth larvae in stored grain
Cotesia glomerata	imported cabbageworm
Diaeretiella rapae	cabbage aphid and other aphids
Macrocentrus ancylivorus	strawberry leafroller and other leafrollers
Macrocentrus grandii	European corn borer
Microctonus aethiopoides	alfalfa weevil
Pholetesor ornigis	spotted tentiform leafminer

Figure 68.
Aphidius smithi,
a braconid wasp
that parasitizes
pea aphid.

Figure 67. The braconid wasp,
Microplitis croceipes, laying eggs
in a corn earworm larva.

Figure 69. The hollow remains, or mummy, of a pea aphid parasitized by the braconid wasp, *Aphidius smithi.* The "trapdoor" was created by the adult wasp at emergence.

Figure 70. *Bathyplectes curculionis*, an ichneumonid wasp that parasitizes alfalfa weevil larvae.

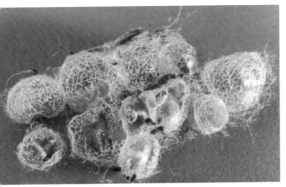

Figure 71. Alfalfa weevil cocoons. The green pupae are alfalfa weevils; the brown pupae are the wasp *Bathyplectes*.

Figure 72. The ichneumonid wasp, *Scambus applopapi*, a parasite of Nantucket pine tip moth larvae.

FAMILY Ichneumonidae: Ichneumonid wasps

Species in U.S. & Canada: 4,780

Size of adults: Usually 3–15 mm, often with conspicuous ovipositor which may be longer than the body (figure 72)

Metamorphosis: Complete—larvae are parasitic

Common hosts/prey: Each species is fairly host-specific—caterpillars, beetle larvae, and sawfly larvae are common hosts, but bee larvae, spiders, insect predators, or even other parasites may also be attacked

Common habitats: Same habitats as their hosts, but more common in moist/humid areas; adults of many species feed on honeydew or nectar

Generations per year: Often 1 (typical life cycle takes 2–4 weeks, often similar to host's life cycle)

Overwintering stage: Often as larvae within their hosts or as pupae

Available for augmentation? *Diadegma insulare* for diamondback moth

Conservation? Reduced use of pesticides; crop diversification and insectary plantings that provide plant foods and alternate hosts

Figure 73. *Diadegma insulare* is an ichneumonid wasp that parasitizes diamondback moth on cabbage.

The Ichneumonidae (figures 70–74) is one of the most diverse families of insects. In addition to the 4,775 named species in the United States and Canada, approximately 3,000 more have been collected but are not yet named or described, and many more probably exist that are as yet undiscovered. All species are parasitic on other insects or related arthropods and are important in the natural control of many insects, but relatively few have been actively used in biological control programs.

Ichneumonids are generally somewhat larger than the closely related braconids. Most species are black, brown, or tan, but some are brightly colored. A few species are capable of stinging humans; however, they are not aggressive or venomous, and incidents of stings are rare.

Like the braconids, some ichneumonid species are solitary, whereas others are gregarious; similarly, the larvae of most species develop inside their host, but some are external parasites.

Table 5. Some species of ichneumonid wasps important in the North Central United States.

Parasite	Pest attacked
Bathyplectes species	alfalfa weevil
Collyria coxator	wheat stem sawfly
Diadegma insulare	diamondback moth
Eriborus terebrans	European corn borer

Figure 74. The brown cocoon of *Diadegma* is formed within the lacy cocoon of its dead host, the diamondback moth caterpillar.

Chalcidoidea: Chalcidoid wasps

Species in U.S. & Canada: 2,220

Size of adults: Very tiny; mostly 0.25–4 mm

Metamorphosis: Complete—larvae of beneficial species are parasitic

Common hosts/prey: Wide variety of hosts; a few are plant feeders, such as the seed chalcids

Common habitats: Same habitats as their hosts; adults of many species use honeydew or flower nectar as a food source

Generations per year: 1 to many (some can complete an entire life cycle in one week)

Overwintering stage: Varies

Available for augmentation? A few of the many available species include: *Trichogramma* species for caterpillar eggs; *Anisopteromalus calandrae* for weevils in stored grains; *Aphelinus abdominalis* for aphids; *Encarsia* species and *Eretmocerus californicus* for whiteflies; *Diglyphus isaea* for agromyzid leafminers; *Thripobius semiluteus* for thrips; *Pseudaphycus angelicus* for certain mealybugs; *Anagrus epos* for certain leafhopper eggs; *Anaphes iole* for lygus bug eggs; *Muscidifurax* species, *Nasonia vitripennis,* and *Spalangia* species for filth fly pupae; *Pediobius foveolatus* for Mexican bean beetle

Conservation? Reduced use of pesticides; crop diversification and insectary plantings that provide plant foods and alternate hosts

The superfamily Chalcidoidea includes about a dozen families of tiny insects. In addition to the known species, hundreds more probably remain unknown to science.

The families of chalcidoid wasps vary considerably in appearance. Color ranges from pale yellow to black; some species are iridescent blue, green, or copper. Because of their diversity and small size, identifying the various families and species requires considerable training.

Most chalcidoid species are parasitic on other insects, attacking most major groups of insect pests. Some chalcidoid families are highly specialized as to the group or stage of host attacked. As an extreme example, all members of the small family Eucharitidae parasitize only ant pupae. All species of the families Mymaridae and Trichogrammatidae parasitize only the egg stage of their hosts. These tiny parasites (adults may be only $^1/_{50}$ inch long!) lay one or more eggs within the eggs of their host insects. The parasite egg hatches, and the larva consumes the contents of the host egg. The parasite then pupates within the host egg, and a new adult parasite emerges to seek out new host eggs.

Figure 75. *Pteromalus puparum* **is a small chalcidoid wasp that parasitizes the pupae of imported cabbageworm.**

In addition to parasitizing their hosts, many adult chalcidoid wasps feed on some of their hosts. This host feeding provides additional nourishment to the adult female wasp, allowing her to produce more offspring for a longer time. One adult can kill several hundred hosts by host feeding, in addition to those killed by its offspring.

A few important groups of chalcidoids are not parasitic. The tiny fig wasp (*Blastophaga psenes*) is an essential pollinator of certain varieties of figs; the wasp larvae live within the fig flowers. Some members of the family Eurytomidae are called seed chalcids because their larvae feed on seeds of plants; this family includes the clover seed chalcid and the alfalfa seed chalcid, both important pests.

Native chalcidoid wasps provide significant natural control of many pests and potential pests. The chalcidoids also constitute the most important group of natural enemies in biological control programs. Many species have been imported for classical biological control, and several species are available commercially for augmentative releases. The following six chalcidoid families are particularly important: Pteromalidae, Chalcididae, Encyrtidae, Eulophidae, Mymaridae, and Trichogrammatidae. Each is discussed below.

The family Pteromalidae (400 species) is one of the most diverse families in the Chalcidoidea. Some species are external, gregarious parasites of the larvae and pupae of beetles and moths. *Pteromalus puparum* (figures 75–76) is a common and important internal gregarious parasite of the pupae of imported cabbageworm. Other species attack the immature stages of flies and wasps. Several species attack house flies, stable flies, blow flies, and other filth-breeding flies (figure 77), and some of these are available commercially.

Figure 76. *Pteromalus* **larvae within the dead pupa of its imported cabbageworm host.**

Figure 77. **The pteromalid wasp,** *Nasonia vitripennis*, **laying eggs into a filth fly pupa.**

Figure 78. *Brachymeria intermedia* is a chalcid wasp that parasitizes gypsy moth pupae.

Figure 80. *Copidosoma floridanum* adults emerge from the mummified cabbage looper host.

Figure 79. The encyrtid wasp *Copidosoma floridanum*, laying eggs into a cabbage looper egg. As the host egg hatches, each *Copidosoma* egg divides many times and can ultimately produce 1000 or more parasite larvae.

Figure 83. The eulophid wasp, *Tetrastichus julis*, laying an egg into a cereal leaf beetle larva.

Figure 81. The encyrtid wasp *Encarsia formosa*, a commercially available parasite of greenhouse whitefly.

Figure 82. Whitefly nymphs parasitized by *Encarsia formosa*. The pale yellow nymphs are unparasitized; the black nymphs are parasitized; the black nymphs with holes were previously parasitized and the adult wasps have emerged.

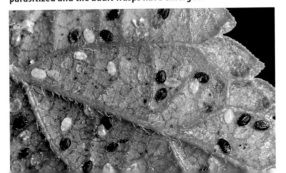

The family Chalcididae (100 species) includes some of the largest chalcidoid wasps; some species are over 10 mm long. These chalcids attack the larvae and pupae of moths, beetles, and flies. *Brachymeria intermedia* (figure 78) is an introduced parasite of gypsy moth pupae.

The family Encyrtidae (500 species) contains many important parasites of ticks, aphids, scale insects, whiteflies, and other insects. One introduced species, *Ooencyrtus kuwanai*, attacks the eggs of gypsy moth. Some encyrtids undergo polyembryony (literally, "many embryos"), in which several offspring develop from a single egg. For example, *Copidosoma floridanum* (figures 79–80) inserts its egg into the egg of a caterpillar such as cabbage looper or soybean looper. After the host egg hatches,

each *Copidosoma* egg divides many times and can give rise to over 1000 larvae. These completely fill the body of the host larva, eventually killing it. One subgroup of this family is sometimes considered a separate family, the Aphelinidae. Many aphelinid species are important parasites of aphids, scale insects, mealybugs, and whiteflies, and several of these have been used successfully in classical biological control. Other species are available commercially for augmentation, notably *Encarsia formosa*, a parasite of greenhouse whitefly (figures 81–82).

The family Eulophidae (500 species) has diverse habits and hosts (figure 83). Few have been used in biological control, but many are important in natural control. Several are parasitic on leafminer larvae and help con-

Table 6. Some species of chalcid wasps important in the North Central United States, grouped by family.

Parasite	Pest(s) attacked
Pteromalidae	
Muscidifurax species	Fly pupae
Nasonia species	Fly pupae
Pteromalus puparum	Imported cabbageworm pupae; other butterfly pupae
Spalangia species	Fly pupae
Eupelmidae	
Eupelmus allynii	Hessian fly and others
Encyrtidae	
Aphelinus mali	Woolly apple aphid
Aphytis mytilaspidis	Oystershell scale
Encarsia formosa	Greenhouse whitefly
Prospaltella perniciosi	San Jose scale
Eulophidae	
Chrysocharis laricinellae	Larch casebearer
Pediobius foveolatus	Mexican bean beetle
Tetrastichus asparagi	Asparagus beetle
Tetrastichus incertus	Alfalfa weevil
Trichogrammatidae	
Trichogramma species	Eggs of several moth species
Mymaridae	
Anaphes luna	Alfalfa weevil eggs

trol such pests as spotted tentiform leafminer on apple and serpentine leafminers on various vegetable and greenhouse crops. Many leafminers that feed on forest trees and landscape plants are attacked by eulophids. Several species are commercially available.

All species in the families Mymaridae (120 species) and Trichogrammatidae (60 species) are parasites of the eggs of other insects. They are tiny, mostly under 1 mm long. Mymarids (figure 84) mostly attack the eggs of leafhoppers and planthoppers, although some species attack eggs of other types of insects. Trichogrammatids, on the other hand, mostly attack the eggs of moths and butterflies, but they can also attack eggs of beetles, true bugs, and flies. *Trichogramma* species (figure 85) have been the most widely used group of natural enemies in augmentation programs worldwide, and many are commercially available from several suppliers in the United States. They are most frequently used for controlling various moth species.

Figure 84. A fairyfly, *Anaphes flavipes* (family Mymaridae), laying eggs into the eggs of cereal leaf beetle.

Figure 85. A *Trichogramma minutum* wasp parasitizing a moth egg.

Additional families of parasitic wasps

Several additional groups of wasps are occasionally important in natural and biological control, but these are generally less common and less important than the wasps described above. These include:

Bethylidae (200 species): Parasitize larvae of wood-boring beetles, moth larvae, and pests of stored grain. *Laelius pedatus* (figures 86–87) shows promise for augmentative releases against pests of stored products, but is not yet commercially available.

Chrysididae: cuckoo wasps (230 species): Parasitize walkingstick eggs and bee, wasp, and sawfly larvae; some kill bee or wasp larvae then feed on the stored food in the nest.

Figure 86. The bethylid wasp, *Laelius pedatus*, laying eggs on a warehouse beetle larva, a pest of stored grain.

Figure 87. The warehouse beetle larva on the right is parasitized by five larvae of *Laelius pedatus*.

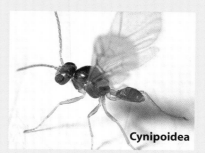

Cynipoidea (superfamily—820 species): Some species parasitize larvae of flies, lacewings, and other wasps; some are hyperparasites; some parasitize cockroach eggs; others are gall-making plant feeders.

Scelionidae (275 species): Parasitize eggs of various pests, including squash bugs, stink bugs, grasshoppers, and tent caterpillars, but some also attack spider eggs.

Scoliidae (22 species): Parasitize white grubs, especially larvae of May and June beetles in the genus *Phyllophaga*.

Tiphiidae (225 species): Some species parasitize beetle larvae in soil, including several white grub species. *Tiphia popilliavora* and *T. vernalis* were introduced into the United States for control of Japanese beetle.

FAMILIES — Vespidae and Sphecidae: Stinging wasps

The two major families of stinging wasps are the Vespidae (420 species), which include yellowjackets, hornets, and paper wasps (figure 88), and the Sphecidae (1,140 species), which include mud daubers, thread-waisted wasps, digger wasps, and cicada killers (figures 89–90). Most of these wasps construct a cell or a nest or dig a tunnel in the soil, within which they lay their eggs. They provision the nests with food for their larvae to eat, usually other insects, often pest species. Some stinging wasps provision their nests with spiders, and these wasps may be detrimental to natural control. The social wasps (yellowjackets and hornets) live in large colonies and care for their young in a fashion similar to honey bees. It has been found that artificially increasing the number of nest locations of *Polistes* wasps (figure 88) will encourage more colonies to develop in an area, providing better biological control.

FAMILY — Formicidae: Ants

The ants are a diverse group of about 700 species in the United States and Canada. Ants are often considered pests because they invade homes, destroy wood, and interfere with outdoor activities. Many can sting and bite. However, most ant species help recycle nutrients and aerate the soil, and many species are predatory on other insects (figure 91). There have been no modern attempts to use ants in biological control programs, although several studies have highlighted their importance in natural control. In the southern United States, fire ants can be the most abundant predators in cotton, corn, and other field crops.

Some ants can interfere with natural or biological control, especially those species that "farm" aphids, mealybugs, and scale insects. These ants "harvest" the sweet honeydew produced by such sap-sucking insects. The ants actually relocate the pests on the plant to start new colonies, and they protect their "herds" from predatory and parasitic insects.

CLASS — Arachnida: Arachnids

ORDERS

- **Araneida (spiders)**
- **Opiliones (harvestmen)**
- **Acari (mites)**

Spiders, mites, scorpions, and several related groups are not insects, which are in the arthropod class Insecta, but belong to the class Arachnida. The spiders (order Araneida) and mites (order Acari) are the two largest groups of arachnids. All spiders are predatory, mostly on insects. Mites are extremely diverse in their habits, ranging from scavengers to plant pests to animal parasites (such as ticks) to predators of other arthropods. Most mites are less than 1 mm long when fully grown.

Figure 91. The ant, *Formica neogagates*, preys on the larvae of gypsy moth and other forest insects.

Figure 88. *Polistes fuscatus* is a common paper wasp in the Upper Midwest. Wasps from the worker caste kill various caterpillars and take them back to the nest to feed the developing larvae.

Figure 90. *Tachysphex* is a sphecid wasp that uses grasshoppers as food for its larvae.

Figure 89. Thread-waisted wasps of the genus *Ammophila* (family Sphecidae) provision their nests with caterpillars.

ORDER ▶ Araneida: Spiders

Species in U.S. & Canada: 3,700

Size of adults: 3–40 mm

Metamorphosis: Simple—both immatures and adults are predatory

Common hosts/prey: Generalist predators, but prey types vary by hunting strategy

Common habitats: Varies by type of spider: soil, low vegetation, or woody plants

Generations per year: Usually 1

Overwintering stage: Eggs or adults

Available for augmentation? No

Conservation? Reduced use of pesticides; reduced tillage; crop diversification and insectary plantings that provide shelter and alternate prey

About 15 of the 68 families of spiders in North America contain species that are commonly found in crop settings. Virtually all species are predators, and most feed on insects.

Spiders capture their prey in three main ways. The most common method makes use of a web. Common families of web spinners include the orb spiders (families Araneidae and Tetragnathidae, figures 92–93), the sheetweb spiders (family Linyphiidae), the combfooted spiders (family Therediidae, figure 94) that construct a very haphazard web, and the funnel web spiders (family Agelenidae, figure 95). The hunting spiders, the second group, do not construct a web to capture their prey, although they may construct a silken refuge. These spiders are very active and often run down their prey to capture it. Examples of hunting spiders include the wolf spiders (family Lycosidae, figure 96), the jumping spiders (family Salticidae, figure 97), the lynx spiders (family Oxyopidae, figure 98),

Figure 95. The funnel-web spiders, family Agelenidae, build a dense funnel-shaped sheet web which leads to a tubular retreat.

Figure 96. Wolf spiders are hunting spiders in the family Lycosidae.

Figure 97. Jumping spiders, family Salticidae, have large eyes to better see their prey.

Figure 92. Most orb-weaving spiders are in the family Araneidae.

Figure 94. The comb-footed spiders, family Therediidae, produce an irregular web for prey capture.

Figure 98. Lynx spiders, family Oxyopidae, are hunting spiders often found in vegetation.

Figure 93. Four-jawed or long-jawed spiders, family Tetragnathidae, also produce orb webs.

and the two-clawed hunting spiders (family Clubionidae, figure 99). The crab spiders (family Thomisidae, figure 100), use the ambush method of prey capture. These spiders are common on flowers and vegetation and sit motionless until their prey comes within easy grasp.

An insect in virtually any active stage of its life cycle can fall prey to a spider. Some spiders will even eat insect eggs or pupae. However, each spider species is more likely to catch a certain prey type, based on its method of prey capture and preferred habitat. For example, orb spiders are more likely to capture adult flying insects than crawling insects. Spiders also reduce pest damage indirectly when they startle potential prey, which then stop feeding and flee.

Spiders can help regulate pest populations, but several attributes may reduce their overall impact as biological control agents. Spiders are generalist predators and do not discriminate among types of prey. A spider is likely to eat anything that is of the appropriate size, occurs in its habitat, and can be captured by its specific prey-capture method. If a particular pest species is abundant, then spiders are likely to prey on it. However, if other prey are more abundant than the pest, most spiders will ignore the pest and feed on the nonpest species. Also, spiders tend to have a single generation per year. Therefore, they are unable to increase their numbers rapidly in response to the buildup of a multi-generational pest.

Many spiders that overwinter as eggs disperse in the spring by ballooning: a newly hatched spiderling spins a thread of silk that catches the wind and carries the spiderling off to a new location. If the habitat where it lands is suitable (adequate humidity, moderated temperatures, and prey), the spiderling will stay there. Otherwise, it may continue to move by ballooning.

Some biological control researchers have questioned the importance of spiders in biological control, whereas others argue that the limited evidence of their importance is not because they are unimportant but because they have been so poorly studied in agricultural systems. However, we do know that spiders are part of the overall natural enemy complex that adds stability and helps keep pest populations from rapidly expanding. No single generalist predator species can provide this buffering; a diverse array of natural enemies, including spiders, is necessary.

The diversity of spiders is greater in undisturbed natural environments than in agricultural settings. In agriculture, spider diversity tends to be greater in less disturbed perennial crops (forests, orchards, vineyards) than in annual crops. Since most spiders seek undisturbed habitats for overwintering, field size may impact spider diversity and population sizes in annual crops: spiders are less likely to recolonize large fields than small fields surrounded by relatively undisturbed vegetation.

Broad-spectrum insecticides are as damaging to spider populations as they are to beneficial insects. Insecticides not only kill spiders directly, but they also kill many nonpest insects that the spiders eat when pests are scarce. Therefore, use insecticides only when needed, and use the most selective materials possible.

Many cultural crop practices also disrupt spider activity. Harvesting and tillage in particular are damaging. When conducted in late summer or fall, these practices can destroy much of the spider population that would otherwise overwinter.

No spiders have been successfully used in natural enemy importation programs. However, human manipulation of spider populations can improve natural control. In China, bundles of straw are placed in fields with high spider populations. The straw provides refuges, and the bundles can be moved to relocate the spiders to other areas. Some research has found that plots covered with straw mulch or overwintered plant cover early in the season help to retain ballooning spiderlings, resulting in larger spider populations and lower pest populations throughout the growing season. In Swiss apple orchards, strips of flowering plants support the webs of orb web spiders, which capture enough winged aphids to significantly reduce aphid populations on the trees.

Figure 99. A two-clawed hunting spider, family Clubionidae.

Figure 100. Many crab spiders, family Thomisidae, are ambush predators; they are frequently found at flowers.

<ORDER> **Opiliones: Harvestmen**

The order Opiliones, also known as Phalangida, contains harvestmen and daddylonglegs, longlegged arachnids that somewhat resemble spiders but are easily differentiated from them. Harvestmen have an oval body (figure 101) whereas spiders have a narrow waist between the cephalothorax and abdomen. In addition, harvestmen are unable to spin silk, nor do they use venom to capture prey. Finally, spiders are exclusively predatory, whereas harvestmen feed on small insects as well as various plants, fungi, and decaying organic matter. Harvestmen are more closely related to mites than they are to spiders. There are about 6400 species worldwide. The name daddylonglegs can be confusing because it also applies to members of the cellar spider family, Pholcidae, which do spin webs.

<ORDER> **Acari: Mites**

FAMILIES

- **Phytoseiidae (predatory mites)**
- **Other families**

Mites are easily overlooked because of their small size. As a group they are almost as diverse as insects, and some scientists believe that there may actually be more species of mites than insects. Many of the 350 families of mites have predatory members. One group used for biological control of fungus gnats and thrips, particularly in greenhouses, are members of the genus *Hypoaspis* (family Laelapidae). But of greatest importance to biological control is the family Phytoseiidae.

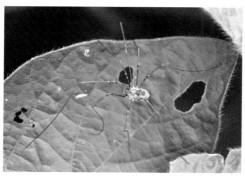

Figure 101. Harvestmen superficially look like spiders, but are in an entirely different arachnid order, the Opiliones.

Figure 102. A predatory phytoseiid mite, *Amblyseius californicus*.

Figure 103. The phytoseiid mite *Phytoseiulus persimilis* is commonly used for augmentation biological control in greenhouses.

<FAMILY> **Phytoseiidae: Predatory mites**

Species worldwide: 2,250

Size of adults: < 1mm (near microscopic)

Metamorphosis: Simple (egg, six-legged larva, eight-legged nymph, adult) —all active stages are predatory

Common hosts/prey: Mites, tiny insects such as thrips and scale crawlers, insect eggs, nematodes, other tiny animals; many can subsist on non-prey food such as pollen

Common habitats: Soil, leaf litter, plants infested with plant-feeding mites

Generations per year: Many; a typical life cycle can take less than a week

Overwintering stage: Varies by species

Available for augmentation? Several species are available, primarily for control of spider mites and thrips. Selection of species depends on type of pest, type of crop, and crop environment. Different species have different tolerances to heat and humidity. *Phytoseiulus persimilis* is probably the most widely used.

Conservation? Pesticide reduction is most important, but dust reduction and management of cover crops or orchard floor vegetation can also be important

Most all phytoseiids are predatory on other mites and small insects. Most live in vegetation, where they are important predators of spider mites (such as European red mite and twospotted spider mite), other plant-feeding mites, and small insects such as thrips, whiteflies, and the crawler stage of scale insects. Many species can also feed on materials such as honeydew and pollen, which allow them to survive when prey is scarce. Although many phytoseiids are generalist predators, several are highly specialized. For example, members of the genus *Phytoseiulus* (figure 103) feed only on spider mites.

Phytoseiids have a short generation time of about 1 week under ideal conditions, and each adult female can produce 40–60 off-spring. During its development, a phytoseiid can consume as many as 20 spider mites, and egg-laying adult females can consume 10 or more prey per day for as long as 2–3 weeks.

Phytoseiids are about the same size as their spider mite prey, about 0.5–0.8 mm in size. They are therefore easily overlooked. However, with some training and good eyesight, it is possible to detect and count phytoseiids in the field. In some cases pest management decisions can be based on the ratio between pest mites and phytoseiids. This works particularly well in apple orchards, where phytoseiids are important natural enemies of European red mite and other pest mites.

Non-native phytoseiid species have occasionally been introduced into new areas for biological control, but no such cases have occurred in the North Central states. Conservation of phytoseiids is an important form of spider mite pest management in orchards and vineyards. In many locations phytoseiids have developed resistance to broad-spectrum organophosphate insecticides, which can then be used for controlling insect pests without interfering with the predatory mites. Phytoseiids have not developed natural resistance to other insecticide groups such as carbamates or synthetic pyrethroids, although artificial resistance to these groups has been induced under laboratory conditions. Several species of phytoseiid mites are raised commercially for augmentative releases (see details on previous page and in chapter 9). However, these tend to be expensive and are used primarily in high-value horticultural crops, such as strawberries and greenhouse crops. North Central Regional publication, *Biological Control of Insects and Other Pests of Greenhouse Crops* (NCR581), contains detailed information on using predatory mites in greenhouses.

Insect-parasitic nematodes

Nematodes, also called roundworms, are a varied group of mostly small animals occupying diverse habits. Many are scavengers, some are fungus-feeders, many are plant-parasitic, and others parasitize various types of animals. Generally, nematodes that parasitize insects do not attack plants or other types of animals, including humans.

Approximately 20 families of nematodes have insect-parasitic species. Nematodes can attack species within most orders of insects. The most common hosts are from the orders Coleoptera (beetles), Diptera (flies, mosquitoes, and their relatives), Orthoptera (grasshoppers, crickets, and their relatives), Lepidoptera (butterflies and moths), and Hymenoptera (sawflies, bees, wasps, and their relatives). In those insects with complete metamorphosis the larval stages usually are parasitized, but in some cases pupae and adults can be attacked. Most insect-parasitic nematodes are slender and only a few millimeters long, but some types can be as long as several centimeters.

Nematodes are essentially aquatic animals; their environment must be moist in order for them to move and survive. Therefore, they are more common in water and moist soil than in drier, more exposed areas. Aquatic species of insect-parasitic nematodes parasitize various aquatic insects, including pests such as mosquito larvae (figure 104). Terrestrial species of insect-parasitic nematodes live primarily in the soil, where they parasitize insect pests such as white grubs, root weevils, wireworms, cutworms, fungus gnats, and maggots. Many plant-feeding insects pupate or spend dormant periods in the soil, and these, too, can be parasitized by nematodes. Some terrestrial nematodes briefly venture into exposed areas to lay eggs or find hosts, but only if there is at least a film of water for their movement, such as after a rain or during a morning dew.

Insect-parasitic nematodes generally have both parasitic and free-living periods. They differ from microbial pathogens in that they are capable of considerable directed movement. Most insect-parasitic nematodes are "cruisers" that actively search out their hosts. A few species, including *Steinernema carpocapsae*, are ambush hunters that move to the soil surface and stand vertically, attached to a soil particle by their tails. When a host passes by, the nematode attaches to the host, even jumping a short distance if necessary.

Figure 104. A mosquito larva with an insect-parasitic nematode coiled in its thorax. The larva will be killed when the nematode emerges.

Two important families of insect-parasitic nematodes are the Steinernematidae and Heterorhabditidae. Members of these two families carry mutualistic bacteria in their digestive systems. These bacteria are themselves insect pathogens and are essential to the nematodes for successful parasitism; therefore, the nematodes in these two families are called entomopathogenic. A generalized life cycle can be summarized as follows.

A juvenile nematode in the infective or parasitic stage will locate a suitable host and enter the insect through its mouth, anus, respiratory openings, or by directly penetrating the cuticle. After the infective juvenile enters the host, it releases from its digestive system the **symbiotic** bacteria that are pathogenic to the insect. The bacteria produce a toxin that kills insect cells. The bacteria multiply and the insect dies, usually within 1–2 days of being parasitized. The infective-stage nematodes grow to adulthood while feeding on the bacteria and/or the dead insect tissues (figure 105). The adults mate and lay eggs within the host insect. There may be three to four continuous generations within the host, as long as there is still acceptable host tissue for feeding. Eventually, juveniles of the appropriate stage will exit the host, carrying with them the symbiotic bacteria, and seek new hosts to parasitize. A large host can give rise to thousands of infective juvenile nematodes (figure 106).

Because of their small size and hidden nature, the benefits of naturally occurring insect-parasitic nematodes are not always well understood. Their benefit in the natural control of plant pests is greatest in areas of continuous moisture rather than in more arid areas. Even in moist situations, however, they may not be abundant enough to provide significant pest suppression. Eight species have been developed commercially for use in augmentation programs (see chapter 9).

Insect pathogens

Many microbial organisms can infect and cause disease in insects. These insect pathogens include viruses, bacteria, fungi, and protists. Many rapidly kill their host insects, whereas others result in more chronic, long-term diseases that can prolong insect development times and reduce reproduction. Some insect pathogens are highly host specific, although some can kill beneficial insects; most are neither pathogenic nor toxic to plants, humans, and other animals.

Figure 105. Adult nematodes (*Heterorhabditis* species) look like spaghetti inside a dead strawberry root weevil larva.

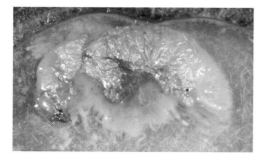

Figure 106. Tiny, slender nematodes spill from the dissected strawberry root weevil larva. Look for the tiny white nematodes near the edges of the photograph.

Insect pathogens can be important in the natural control of pest populations. Some diseases can rapidly spread through an insect population and cause high mortality in a relatively short period of time. This phenomenon, called an **epizootic**, is equivalent to a rapid disease epidemic in a human population. Epizootics usually occur only at high insect population densities, which enable the pathogens to spread from insect to insect. For this reason, you usually cannot wait for an epizootic to control a pest of a high-value crop, because serious damage can occur as the pest population is increasing. However, epizootics are very important in the natural control of pests of forests, rangeland, and certain field crops.

Some insect pathogens require specific environmental conditions before they can infect and cause disease, which can limit their effectiveness in controlling pests. For example, insect-pathogenic fungi require a relatively high humidity for the fungal spores to germinate and infect the host insect.

A growing number of insect pathogens are produced commercially for application through traditional insecticide-application equipment. The use of such microbial insecticides is called **microbial control**. This form of natural enemy augmentation is discussed in more detail in chapter 9. There is also a growing number of pesticides that do not contain viable insect pathogens but whose active ingredients are derived from microorganisms. Use of these microbially derived insecticides is more like pesticidal control than biological control, since live natural enemies are not involved.

As with predatory and parasitic insects, it is necessary to know some characteristics and attributes of the insect pathogens that are important natural or biological controls in your crops. The following discussion provides a general background and a few important examples.

Insect-pathogenic viruses

Viruses are submicroscopic particles that can only reproduce within living cells of a host organism. Because of this intimate relationship between a virus and its host, few viruses are able to attack more than a single species of host. Larvae of moths and butterflies are the most common hosts of insect-pathogenic viruses, but other important host groups include the sawflies, true flies, beetles, grasshoppers, true bugs, and sap-feeding insects such as aphids, scale insects, leafhoppers, and related insects.

Viruses usually do not have latinized names like those used for other types of organisms. Instead, they are usually named based on their general group and their host insect. Important groups of insect viruses are the nuclear polyhedrosis viruses (NPV), cytoplasmic polyhedrosis viruses (CPV), and granulosis viruses (GV). Therefore, the nuclear polyhedrosis virus that attacks the corn earworm, *Helicoverpa zea*, is often simply called *Helicoverpa* NPV, and the granulosis virus that infects codling moth larvae is called codling moth GV.

Three families of insect-pathogenic viruses (including the families of the CPVs, GVs, and NPVs) enclose their infective virus particles (virions) in tough survival packages called **occlusion bodies**. An occlusion body can protect its virions from ultraviolet light and other adverse environmental conditions for up to several years.

Insect viruses usually have to be eaten by a suitable host in order to cause infection, although some can be transferred from parent insect to offspring through the egg. We can use a generalized NPV as an example of the infection and multiplication process. Polyhedra—a type of occlusion body containing many virions—are ingested by a larva during feeding and pass to the insect's midgut, where the polyhedra dissolve and release their virions. The virions pass into the insect's body and invade susceptible cells. The virions multiply, the host cells burst, and the newly produced virions attack other cells. As this process continues, virions begin to congregate, forming new polyhedra. Eventually, all of the susceptible cells are killed, as is the host insect, and millions of polyhedra have been formed. As the dead insect decomposes, it liberates the polyhedra into the environment, where they may be consumed by other susceptible hosts.

The NPVs and GVs tend to be the most virulent of the insect viruses, capable of causing rapid development of symptoms and a high percentage of deaths. The incubation period (from ingestion to the start of symptoms) is a few days to a week. Death follows a day or two later. Symptoms include sluggishness, cessation of feeding, and pale discoloration. At death, larvae become limp and often hang from their **prolegs**, either head-down or in an upside-down V (figures 107–108). The exoskeleton becomes shiny and often is fragile and easily broken.

Wind and rain are important in dispersing insect-pathogenic viruses. In some hosts, the virus manipulates the behavior of the host, causing it to climb to a high location and grab on shortly before it dies. This permits increased dispersal of the virus. Birds, small mammals, and insect parasites also spread insect-pathogenic viruses when they feed on or oviposit into infected insects.

Viruses cannot reproduce unless they are within a cell of a suitable host, so commercial production of a virus requires maintenance of a culture of the host organism. This makes it fairly expensive to mass produce insect-pathogenic viruses. In addition, these viruses have a limited market because they are specific to one or a few hosts, remain effective in the field for a relatively short time, require over a week to kill their hosts, and have a limited shelf-life. Despite these problems, several insect-pathogenic viruses have been produced commercially for use in the United States, including codling moth granulosis virus and NPVs for alfalfa looper, beet armyworm, Douglas fir tussock moth, gypsy moth, tobacco budworm, and pine sawflies. In Brazil, velvetbean caterpillar (*Anticarsia gemmatalis*) NPV is used on approximately 2.5 million acres annually.

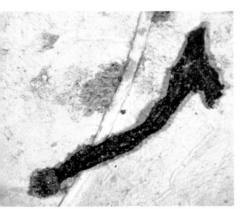

Figure 107. This cabbage looper was killed by a nuclear polyhedrosis virus. A dead larva hanging from its abdominal prolegs is a characteristic symptom of a virus infection.

Figure 108. A gypsy moth larva killed by a nuclear polyhedrosis virus.

Insect-pathogenic bacteria

Bacteria are microscopic single-celled organisms. Though much larger than viruses, it might take as many as 100 of them lined up end-to-end to cross the period at the end of this sentence. Many species of bacteria cause insect diseases, but relatively few are important in pest control. The most important are in the genus *Bacillus*. Some species can multiply only in host insects, whereas other species can also multiply outside of their host under certain conditions. When conditions become unfavorable, most *Bacillus* species form resilient spores that can survive adverse environmental conditions. Insect-pathogenic *Bacillus* species also produce protein toxins that harm the insect hosts. Some of these toxins are produced as solid structures, called **parasporal bodies**, sometimes called crystals, during the spore-formation process.

Typically, insects must ingest the bacteria or their spores in order to become infected, and early instars (younger larvae) are usually much more susceptible than later instars. Symptoms of bacterial infections are fairly generalized and include loss of appetite, sluggishness, discharge from mouth and anus, general discoloration, and eventually liquefaction and putrefaction of most body tissues. The insect exoskeleton is not usually affected, as it is with some of the insect viruses.

Bacillus thuringiensis (Bt) is the most important bacterium in pest control. During spore formation, Bt produces a polyhedral-shaped crystal parasporal body (figure 109) which is toxic to susceptible insects. At least 60 different **subspecies** have been named and many thousands of strains have been identified. There is great variability in the toxins produced by these strains and subspecies and, consequently, considerable variability in their host ranges. Bt is most familiar as a pathogen of caterpillars, but strains of Bt have been isolated that are pathogenic to bees and wasps, beetles, fly larvae, true bugs, chewing lice, and even some nematodes and protists.

The typical progression of disease in a caterpillar that has ingested Bt is as follows. The caterpillar ingests spores and crystalline parasporal bodies during feeding. In the midgut, the crystals dissolve, releasing a toxin that damages the cells lining the gut. Gut paralysis occurs within a few minutes, and feeding stops. Leakage of stomach contents through the midgut lining allows spores to enter the insect's blood cavity, where the spores germinate. The bacteria attack many of the internal tissues and multiply until conditions are no longer favorable, at which time the insect's normal body color has changed (figure 110) and new spores and parasporal bodies are formed. Discharge from the mouth and anus, as well as the eventual deterioration of the cadaver release spores into the environment, where they can infect additional hosts. However, disease transmission from infected to uninfected hosts is less common with Bt than with other types of insect pathogens.

Bt can survive and reproduce in soil under favorable conditions. However, the spores rapidly lose their viability if exposed to bright sunlight and dry conditions. Several companies mass-produce strains of Bt for sale as microbial insecticides; specific strains are available for specific pests, such as caterpillars, certain beetles including the Colorado potato beetle, and mosquitoes and fungus gnats (see chapter 9).

In some cases, the parasporal toxin of a Bt strain by itself, without infective spores, is sufficient to cause gut paralysis and death in susceptible insects. The genes for such toxins have been successfully implanted into many crop plants through genetic engineering, so that the toxin is synthesized in high doses throughout the plant. Susceptible insects that feed on these genetically modified plants quickly stop feeding and die.

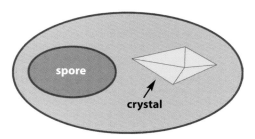

Figure 109. A schematic diagram of a sporulating cell of *Bacillus thuringiensis*.

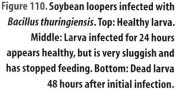

Figure 110. Soybean loopers infected with *Bacillus thuringiensis*. Top: Healthy larva. Middle: Larva infected for 24 hours appears healthy, but is very sluggish and has stopped feeding. Bottom: Dead larva 48 hours after initial infection.

Because these toxins are produced throughout the plant and throughout the growing season, a major concern is that pest populations feeding on these plants could develop resistance to the toxin. Should this happen, they would be resistant not only to the Bt toxin as found in the plants, but also to the same strain of Bt applied as a microbial insecticide. Several strategies to prevent this have been proposed, but only one has been adopted so far. Growers who plant genetically modified crops are required to plant unmodified, unsprayed areas of the same crop nearby to serve as refuges for insect pests that are susceptible to Bt. The goal is to allow susceptible insects from the refuge areas to mate with any insects that develop resistance, resulting in offspring that are susceptible to Bt, thereby slowing or stopping the spread of resistance.

Thus far, field studies have revealed minimal adverse effects of genetically modified Bt-producing crops on natural enemy diversity or abundance, honey bees and other nontarget insects, or soil biology.

About a half dozen species of *Paenibacillus* (formerly included as members of the genus *Bacillus*) cause "milky diseases" of white grubs. The disease name refers to the white discoloration of the insect blood, which is normally clear. *Paenibacillus popillae* and *P. lentimorbus* cause milky disease of Japanese beetle larvae. They will not form spores unless grown in their insect hosts. However, spores can survive for many years in the soil until they are ingested by grubs as they are feeding. These two species have been commercialized for use as microbial insecticides for control of Japanese beetle grubs, though only *P. popillae* is currently available. In some cases, application of these bacteria during one growing season has been credited with controlling Japanese beetle grubs for many years. Other studies have shown low rates of infection and minimal control of grub infestations (see chapter 9).

Insect-pathogenic fungi

Most fungi are multicellular organisms that grow as threadlike hyphae forming a web-like mass called mycelium. Whereas insects must eat viruses, bacteria, and protists to become infected, spores of insect-pathogenic fungi usually germinate on the outside of the insect's cuticle and their hyphae penetrate through the cuticle. Because they do not need to be ingested to cause infection, fungi are the only important pathogens of sap-feeding insects such as aphids, whiteflies, and leafhoppers. Whatever the host, the infection process requires relatively high humidity, limiting the effectiveness of fungi under dry conditions. However, fungal spores can be highly resistant to environmental extremes, and under ideal conditions some fungi can reproduce and spread rapidly, resulting in fungal epizootics.

The fungal disease cycle can be generalized as follows. A spore lands on the outside of an insect's body. Under favorable temperature and moisture conditions, the spore germinates and the fungal hypha penetrates the cuticle into the insect body. The fungus produces toxins that kill the insect and allow for rapid hyphal growth. Eventually, the entire body cavity is filled with fungal mass. Hyphae then penetrate outward through the softer parts of the insect's body, often "gluing" the insect to the place where it died. Under favorable moisture conditions, the external hyphae will produce spores that ripen and are released into the environment, completing the cycle. Fungal spores can be dispersed by water or wind, or on the surfaces of animals or equipment. Like some viruses, some fungi cause their host to climb to higher locations and grab on, late in the infection. Fungus-killed insects are often seen clinging to the tip of a branch or leaf, often high up on a plant, allowing more effective dispersal of spores.

Over 700 species of insect-pathogenic fungi have been identified, but the importance of most of these in the natural control of insect populations is unknown. Many genera of fungi attack insects, but the most important are *Metarhizium*, *Beauveria*, *Entomophthora*, *Entomophaga*, and *Zoophthora*. *Metarhizium anisopliae* and *Beauveria bassiana* (figure 111) both have a wide host range that includes such diverse groups as grasshoppers, true bugs, aphids, caterpillars, and beetles. The genus *Entomophthora* (which has recently been reclassified and split into several other genera) consists of several important species. *Entomophthora muscae* is an important species that kills many species of adult flies (figure 112), including pests (such as seed corn maggot) and beneficials (such as hover flies). During periods of humid weather, it is common to find dead flies stuck to window panes, leaves, fence wire, or other surfaces; these have been killed by *E. muscae*. Other important species include *Zoophthora radicans*, which attacks potato leafhopper, and *Z. phytonomi*, which kills alfalfa weevil. Beginning in 1989, epizootics of a fungus called *Entomophaga maimaiga* (figure 113) were observed in gypsy moth infestations in the northeastern United States, and the fungus has gradually spread westward, following the spread of gypsy moth; the origin of the *E. maimaiga* responsible for this series of epizootics is unknown.

Figure 111. Soybean loopers killed by the fungus *Beauveria bassiana*.

Because of their environmental requirements, insect-pathogenic fungi are not consistently reliable except in areas of high humidity. Activity can be enhanced by using them as microbial insecticides (see chapter 9) or by artificially increasing humidity (such as in a greenhouse or with overhead irrigation), but artificially increasing humidity can promote outbreaks of fungal plant pathogens as well. In China, *B. bassiana* has been widely used against corn borers. In Brazil, *M. anisopliae* has been used extensively against sugarcane spittlebugs, because fungal applications are compatible with releases of parasitic insects that target other pests.

Research into fungal insect pathogens continues, and it is likely that additional practical applications will be developed in the future.

Figure 112. A fly killed by the fungus *Entomophthora muscae*. Flies killed by this fungus are commonly seen stuck to vegetation, windows, or other surfaces.

Insect-pathogenic protists

Insect-pathogenic protists are single-celled microbes that differ from bacteria in various details of their cellular structure. They also differ from bacteria in that most insect-pathogenic protists can only grow and reproduce while inside a cell of a host insect.

Although some insect-pathogenic protists rapidly kill their hosts, others cause chronic infections. A protist infection often shortens the life span of the insect and can reduce reproduction. This means that protists often exert a constant, low-level suppression of insect populations rather than the sudden population crash of a rapid epizootic.

There are insect pathogens within many major groups of protists, but the most important group consists of the microsporidians, which are spore-formers; recent studies suggest that these should be classified with the fungi instead of the protists. Two genera are of particular importance. The genus *Nosema* has several species and attacks insects in at least 10 orders, especially the Lepidoptera (caterpillars), Coleoptera (beetles), Orthoptera (grasshoppers, crickets, and their relatives), and Hymenoptera (sawflies, bees, wasps, and ants). *Nosema locustae* infects many species of grasshoppers, katydids, and crickets. After spores are ingested, *N. locustae* rapidly reproduces in fat tissue, causing the abdomen to swell. The infection slows the insect's development and can reduce or eliminate egg production in females and sperm production in males. Females that do lay eggs can pass the microsporidian to the next generation through the eggs. The

death rate from *Nosema* infections is relatively low, but the effect on reproduction reduces the host population in subsequent generations. *Nosema locustae* is commercially available as a microbial insecticide (see chapter 9). Another *Nosema* species, *N. pyraustae*, is occasionally important in the natural control of European corn borer.

Species in the genus *Vairimorpha* attack primarily caterpillars. *Vairimorpha necatrix* is one important species. It is somewhat more virulent than *N. locustae*, resulting in a higher death rate.

Lagenidium giganteum (formerly classified as an oomycete fungus) has been produced commercially and used successfully for control of mosquito larvae, including species in the genera *Aedes* and *Culex*.

Because of their relatively low virulence, and the relatively long infection period preceding death, most protists are more important in natural control than in biological control. Attempts to introduce non-native protists into pest populations have not resulted in spectacular successes, but there is good potential for at least low levels of suppression. All insect-pathogenic protists must be grown in live cells, which makes production of protist-based microbial insecticides expensive. Also, spores tend to die rapidly when exposed to sunlight and dry conditions. Protists will continue to be important in natural control, but applications in biological control will probably remain limited to specific cases.

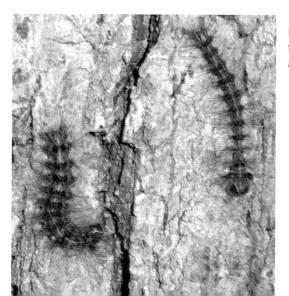

Figure 113. The gypsy moth larva on the right was killed by the fungus *Entomophaga maimaiga*.

Vertebrate predators

Bats, birds, frogs (figure 114), and toads can be important in the natural control of insects, and their contribution to biological control can be enhanced through appropriate conservation practices.

Bats feed at night and capture flying insects such as moths, beetles, stink bugs, leafhoppers, and mosquitoes. Their presence in a given area depends largely on the availability of adequate shelter, but they also prefer access to pools of water at least 10 feet wide from which to drink while on the wing.

Birds can complement the activity of bats by feeding on insects during the day. Some birds feed primarily on insects during the nesting season, then switch to a diet of mostly seeds afterwards. Others will take advantage of cherries, berries, and other fruits, and can cause economic losses if these are being grown as crops. In field settings, bluebirds, barn swallows, wrens, sparrows, and starlings can consume many insects, at least at times. In orchards, titmice, nuthatches, chickadees and woodpeckers are important predators of insects on trees throughout the year, not just during the summer. Because the various insect-eating birds prefer different habitats and forage in different ways, conservation practices (housing, water, supplemental foods) must target the desired species in order to be effective.

The American toad can be found throughout the North Central region, and various other toad and frog species have more limited distributions. All of these consume large numbers of insects, slugs, worms, and other invertebrates. Their presence in agricultural settings depends on a diversified habitat that provides shelter, adequate prey, and pools of water for breeding sites. Some toads can live upwards of 10 years under suitable conditions, so provision of toad-friendly habitat can yield long-lasting enhancement of natural control.

Figure 114. Frogs, lizards, birds, bats, skunks, rodents, and other vertebrates eat lots of insects, but they tend to be generalist predators, and mostly they are difficult to manage for pest control purposes.

The introduction of non-native natural enemies

Many insects are serious pests because they were accidentally introduced into a new area without the natural enemies that control them in their native locations (see page 14 for examples). The most dramatic successes in biological control have resulted from the importation of natural enemies from other countries, a practice called both *classical biological control* and *importation biological control*. The goal is to find useful natural enemies and introduce them into the area of the target pest so that they become permanently established and provide continuing pest control with little or no additional human intervention. The search for natural enemies in other countries is called foreign exploration. The first major success occurred over 100 years ago, when the vedalia beetle (figure 115) was imported from Australia and New Zealand to control cottony cushion scale, a serious pest of the California citrus industry. Since then, over 5,000 introductions of about 2,000 arthropod species worldwide have permanently reduced the populations of at least 165 species of pests.

Introduced natural enemies complement the natural control provided by native natural enemies. In some cases, a single introduced natural enemy has been sufficient to provide outstanding biological control. However, often, two or more introduced natural enemies may be needed to provide the desired level of control. Further, native generalist natural enemies may also work in concert with the new introduced species. For example, six parasitic wasp species were introduced—and have become established in most of the Midwest—for control of alfalfa weevil. Together with various native natural enemies, these wasps adequately control alfalfa weevil in more northern regions of the United States (although alfalfa weevil may still cause economic losses in southern regions unless other control measures are taken). European spruce sawfly is under successful biological control due to the combined effects of an introduced virus and at least seven introduced parasites, and the larch sawfly has come under control due to the combined effects of two introduced parasites, one native parasite, and various native small mammals that prey on sawfly pupae in the soil.

Classical biological control differs from conservation (chapter 8) and augmentation (chapter 9) of natural enemies because it is not directly conducted by the farmer or gardener. International agencies, federal agencies (especially the United States Department of Agriculture), and state agencies (state departments of agriculture and land-grant universities) are responsible for the process. These organizations identify potential target pests, locate their country or region of origin, search these areas for candidate natural enemies, evaluate candidate natural enemies in quarantine facilities, and introduce selected natural enemies into the necessary areas. Indeed, quarantine laws prohibit private in-

Figure 115. The vedalia beetle, *Rodolia cardinalis*, a predator of cottony cushion scale on citrus, and the first successful case of biological control by the importation of a non-native natural enemy.

dividuals or agencies from introducing non-native organisms, including natural enemies, without proper authorization from the USDA. Trained personnel must carefully screen natural enemies under rigid quarantine conditions to be certain that they (1) will provide benefit in controlling the target pest, (2) will not adversely affect species other than the target pest, and (3) do not harbor their own natural enemies that might interfere with their effectiveness.

Classical biological control involves numerous logistical and financial challenges related to exploring foreign lands for natural enemies, getting candidate natural enemies to a quarantine facility alive, and successfully rearing them in quarantine during the evaluation process. As a result, biological control workers often have a limited selection of candidate natural enemies with which to work. However, to the extent possible, natural enemies chosen for use in a classical biological control program should have these attributes:

- have a good ability to seek out and aggregate near the target pest
- have a life cycle well synchronized with that of the target pest
- be well adapted to local climate, including the ability to survive the winter
- have a short development time relative to that of the target pest
- engage in host-feeding, if a parasitic wasp
- disperse well
- attack a limited range of hosts or prey

Many of the successes in classical biological control have occurred in tropical and subtropical locations. California, Hawaii, Texas, and Florida have successfully introduced numerous non-native natural enemies. Each of these states uses state revenues to support biological control efforts. Historically, midwestern states have provided little funding for classical biological control. The few significant cases in the region have resulted primarily from USDA programs. State involvement in, and regional coordination of, classical biological control projects offers two advantages over nationally run programs: (1) state or regional programs can better ensure that selected natural enemies are optimally adapted to local environmental conditions, and (2) they can attempt biological control of minor and localized pests that would be of less interest at a national level.

Most classical biological control programs targeting insect pests have used insect natural enemies. However, other types of natural enemies have been imported and released against insect pests. For example, non-native nuclear polyhedrosis viruses have provided significant control of spruce sawfly in Canada and soybean looper in Louisiana.

Optimum environmental range of natural enemies

All organisms have a certain range of environmental conditions in which they can survive, including availability and quality of food, climatic conditions, and presence of natural enemies and competitors. For an organism to thrive, however, it requires a much narrower range of environmental conditions called its optimal environment. This is as true for natural enemies as it is for all organisms. Given the size and geographic and environmental diversity of the United States, a natural enemy could be introduced into an area with optimal conditions or one with suboptimum conditions. The level of biological control that results will be determined in large part by how well the environmental conditions match the natural enemy's optimal conditions. In suboptimum areas, the natural enemy will provide, at best, only limited or partial control of the target pest. A classical biological control program will be most successful when natural enemies are selected from a region that has a climate that is as similar to that of the release area as possible.

Control of minor and localized pests

Most foreign exploration programs undertaken by federal agencies are for natural enemies of major pests with widespread distribution. However, significant pest problems may occur on minor crops or in limited geographical areas.

State departments of agriculture and land-grant universities in the North Central United States have shown increasing interest in biological control. It is hoped that state (as well as federal) agencies will increasingly support foreign exploration and establishment of non-native natural enemies specific for pests in this region. Table 7 presents some of the partial and substantial successes of classical biological control in the Midwest. The list of introduced pests in chapter 3 (table 2, page 14) includes many additional targets for classical biological control in the future.

Increasing the benefits of non-native natural enemies

Although farmers and gardeners are not directly involved in the implementation of classical biological control, their actions can directly influence the effectiveness of non-native natural enemies that do become established. In fact, once an introduced natural enemy has become established, its ability to suppress pest populations is determined by the same environmental and management processes that affect native natural enemies. Recognizing these natural enemies, understanding their benefits, and knowing how to include them in an overall pest management program are important aspects of both conservation and augmentation of natural enemies, the subjects of the next two chapters.

Table 7. Successful establishments of non-native natural enemies in the North Central United States.

Pest	Crop	Established natural enemies	Type of natural enemy	Source
alfalfa blotch leafminer	alfalfa	*Chrysocharis punctifacies*	larval-pupal parasite	Europe
		Dacnusa dryas	larval-pupal parasite	Europe
alfalfa weevil	alfalfa	*Anaphes luna*	egg parasite	uncertain
		Bathyplectes anurus	larval parasite	Europe
		Bathyplectes curculionis	larval parasite	Italy
		Microctonus aethiopoides	adult parasite	France
		Microctonus colesi	larval-adult parasite	uncertain
cereal leaf beetle	grain crops	*Anaphes flavipes*	egg parasite	W. Europe
		Tetrastichus julis	larval parasite	Europe
clover leaf weevil	clover and alfalfa	*Biolysia tristis*	larval parasite	Italy
elm leaf beetle	elm	*Tetrastichus gallerucae*	egg parasite	Europe
euonymus scale	euonymus	*Chilocorus kuwanae*	predator	Korea
European corn borer	corn	*Macrocentrus grandii*	larval parasite	Europe, Asia
		Eriborus terebrans	larval parasite	Europe, Asia
gypsy moth	trees	Over 10 species	predators, parasites, pathogens	Europe and Japan
imported cabbageworm	crucifers	*Cotesia glomerata*	larval parasite	Europe
Japanese beetle	turf grass	*Tiphia vernalis*	larval parasite	Korea, China
larch casebearer	larch	*Agathis pumila*	larval parasite	England
		Chrysocharis laricinellae	larval parasite	England
Nantucket pine tip moth	pines	*Campoplex frustranae*	larval parasite	Virginia
oriental fruit moth	stone and pome fruits	*Macrocentrus ancylivorus*	larval parasite	New Jersey
Russian wheat aphid	grain crops	*Aphelinus albipodus*[a]	parasite	Eurasia
		Aphidius uzbekistanicus[a]	parasite	Eurasia
		Praon gallicum[a]	parasite	Europe
soybean aphid	soybean	*Binodoxys communis*[b]	parasite	Asia

[a] Originally introduced into the western United States and eventually relocated to the North Central states.
[b] First released in several North Central states in 2007, so level of success unknown at printing.

Soybean aphid biological control: A model for industry–university–USDA partnerships

The soybean aphid, a destructive newcomer to North America, has shown the potential for effective partnerships among key groups to mobilize, research, and implement biological control measures.

Soybean aphid, a native of Asia, was first detected in North America in 2000. By the end of that first year it had been found in 10 states. Within 2 years it had spread to 20 states and was causing significant economic damage.

In response, University researchers in the North Central states rapidly initiated programs to study the pest's biology, extent and types of damage, and management. From the beginning, the soybean industry supported this research through state and regional funding.

An early benefit of this research was the identification of several existing generalist natural enemies, especially lady beetles and pirate bugs, that were providing some population regulation of soybean aphid. But it was also clear that these natural enemies could not maintain aphid populations below damaging levels.

In 2004, research and Extension entomologists from six North Central land grant universities and the USDA's Agricultural Research Service submitted to the North Central Soybean Research Program an integrated 3-year proposal on classical biological control of the new pest. The grant, awarded in 2005, supported several foreign exploration trips to Asia, which led to the identification and collection of a small group of candidate natural enemy species.

Several species were researched under strict quarantine conditions provided by the USDA and participating universities. One parasitic wasp species, *Binodoxys communis*, showed particular promise, leading the USDA and state departments of agriculture to approve release permits. The first releases were made in seven states in 2007. As this publication was going to press in early 2008, at least four other species were looking promising; it appears likely that one or more of these will be approved for release.

Although the final outcome of soybean aphid biological control won't be known for several years, this project serves as a model for other possible classical biological control programs in the North Central states. In partnership with local land grant universities, commodity groups can identify and support, both politically and monetarily, the research that must be done to successfully implement classical biological control on pests affecting their own crops.

A *Binodoxys communis* wasp parasitizes a soybean aphid. First released in 2007, this introduced species shows promise as an effective natural enemy of the aphid.

Protecting natural enemies

In many situations, natural enemies are present but are somehow hampered so that they are not fully effective. It is easy to forget that they live in a world that is larger and more complex than just the pests you want them to attack. Natural enemies interact with many other organisms in their habitats; some are sources of food, some are competitors, some are natural enemies of the natural enemies, and some, especially plants, are major structural components of the habitat. In addition, physical factors such as temperature, wind, humidity, precipitation, and light have significant impacts on the development, survival, reproduction, and movement of natural enemies. When biological or physical conditions become unfavorable, natural enemies may die, languish, or disperse in search of more favorable surroundings, reducing their performance as biological control agents.

The structure of the surrounding landscape can also influence pest abundance and natural enemy effectiveness. In general, natural enemies are more dependent on habitats outside of cropping areas than are pests. In a Michigan study, for example, *Eriborus terebrans* parasitized two to three times more first-generation European corn borers near the edges of fields bordered by woods than in the interiors of the fields, presumably because the woods provided a cool, humid habitat and floral foods for the wasps. Another study found that overall parasitism of true armyworm was over three times higher in complex landscapes than in simpler landscapes. There is still much we do not know about how landscape structure affects natural enemies. However, on a more local scale, appropriate management of pests, natural enemies, crops, and surrounding habitats, may ameliorate the biological or physical factors that reduce natural enemy effectiveness. These practices, called *conservation biological control*, are aimed at maximizing the potential of native or imported natural enemies, as well as those released locally for augmentative biological control (see chapter 9).

Different natural enemies have different needs, and different habitats offer different combinations of resources and adversities, making it impossible to list the specific factors that limit natural enemy effectiveness under all situations. However, some of the more important problems natural enemies face include:

- lack of supplemental foods such as pollen, nectar, and honeydew for adult natural enemies
- lack of hosts or prey
- attack by natural enemies of their own
- habitat characteristics that impede searching for pests
- unfavorable physical conditions (too hot, dry, windy, dusty, etc.)
- lack of suitable sites for overwintering
- direct mortality from pesticides, tillage, and other management practices

Other than minimizing the impacts of insecticide use on natural enemies, diversification of noncrop vegetation has probably been the most widely adopted approach to conserving natural enemies. The presumption is that, compared to monocultures, diverse habitats offer more favorable environmental conditions, more supplemental foods such as pollen, nectar, and honeydew, and more alternate hosts and prey to sustain natural enemies when the target pest is scarce. Most plant pests are more likely to find their host plants and remain on them in simple habitats such as monocultures than when their host plants are embedded in a more complex habitat. Unfortunately, we do not yet know enough to predict with certainty the conditions under which diversification of a specific cropping system will favor natural enemies over pests. Different cropping systems are

Figure 116. Hover fly taking essential nourishment from a flower. The adults of many natural enemies obtain nectar and pollen from flowers.

home to different pests and natural enemies, and the actual species present in a cropping system can vary greatly by location and between years at a particular location. For these reasons, it is impossible to offer a list of approaches to conserving natural enemies that will work everywhere, all the time. Instead, the rest of this chapter offers general guidelines and examples that can guide you in developing management practices that take full advantage of the natural enemies in your particular situation.

Conservation biological control practices, as described in this chapter, fall into three categories: (1) developing appropriate crop management practices, (2) providing required resources for natural enemies, and (3) reducing interference with or mortality of natural enemies. No firm lines separate these categories. In fact, they often interact and overlap. The examples we describe are not all-inclusive, but they demonstrate some of the practices that conserve natural enemies.

Crop management practices

Crop management can affect natural enemies and pests in a variety of ways. The following are a few examples.

Intercropping

Natural enemies may prefer a different habitat from the main crop or may need resources that are unavailable in the main crop. Intercropping, the cultivation of more than one crop in the same place at the same time, can enhance biological control if the adjacent crops offer different resources or environmental conditions for natural enemies. In any form of intercropping, it is important to choose compatible crops, not only for the natural enemies, but for overall farm management. There are several types of intercropping, all of which have been shown to enhance biological control in at least some cases. For more details, see the USDA publication *Intercropping Principles and Production Practices: Agronomy Systems Guide* (attra.ncat.org/attra-pub/PDF/intercrop.pdf).

In *row intercropping*, the main crop is grown in rows and one or more additional crops are grown in the row or underneath; undersowing is a common example.

In *strip intercropping*, different crops are grown in adjacent strips that are wide enough to allow management by separate field equipment but close enough to allow the crops to interact. For example, alfalfa, which harbors many generalist natural enemies, can be planted in alternating strips with other crops such as cotton or corn to enhance the overall habitat for natural enemies.

Mixed intercropping involves the cultivation of different crops with no distinct row arrangement. Except for forages, where mixed-species hayfields generally experience less damage from potato leafhopper than single-species hayfields, mixed intercropping is usually not practical in mechanized agriculture.

In *relay intercropping*, a second crop is planted into a standing crop when the standing crop is well established but not yet ready for harvest. For example, tomatoes can be transplanted into a month-old lettuce crop or soybeans can be seeded into winter wheat in late spring.

Alternate harvesting

Staggering harvest times may conserve natural enemies. Cutting alfalfa provides a good example. When you harvest an entire field simultaneously, natural enemies that survive the harvesting operation tend to leave the area, allowing for resurgence of such pests as aphids, alfalfa caterpillar, and armyworms. If you harvest the field in alternating strips, natural enemies may stay in the field and provide better biological control. This practice is called strip cutting. On-farm research in Iowa showed that simply leaving a single 15-foot wide strip uncut at the time of the first cutting was adequate to retain natural enemies which colonized the regrowth and provided good biological control of alfalfa weevil.

Residue management

Crop management recommendations usually include removal and/or destruction of crop residues to destroy pests and plant pathogens. However, this practice also eliminates natural enemies in those residues. In some cases, the benefits of conserving these natural enemies outweigh the risks associated with not destroying the residues. Compared to conventional tillage, no-till farming systems often support more natural enemies, especially ground beetles, rove beetles, and spiders.

Providing natural enemy requirements

Natural enemies require various resources, such as food, water, and shelter, that may be unavailable or in limited supply in the crop environment. Under such circumstances, natural enemies leave or perish. The following are a few examples of how to supply resources to encourage natural enemies to remain in a given area.

Habitat provision and management

Some natural enemies have specific habitat requirements. For example, predatory social wasps in the genus *Polistes* (paper wasps, figure 88) build nests for their young and provide caterpillars as food for their larvae. Most cropping situations have few available sites for the wasps to build their nests. Providing nest boxes can increase the numbers of these wasps and reduce caterpillar damage to cotton, tobacco, cabbage, and other crops. Similarly, the contributions of bats and birds to biological control can be enhanced by providing appropriate housing and access to water. Because different species have different habitat and housing preferences, it is important to know which species are likely to be important in your situation so you can tailor resources to their needs.

Manure management is important for conserving the natural enemies of filth flies. Different natural enemies have different requirements for the depth, structure, and moisture content of manure. Manure management practices can provide the right habitat conditions to conserve more species of natural enemies.

Farmers in the United Kingdom and several other European countries have constructed permanent raised earthen banks through the centers of fields. These "beetle banks," approximately 1.5 feet high by 6 feet wide, are planted with perennial clump-forming grasses and provide undisturbed overwintering habitat for ground beetles, rove beetles, spiders, and other natural enemies of aphids in small grains. A beetle bank in the center of the field allows natural enemies to more quickly colonize all of the field as they disperse from their overwintering sites in the spring. In 2006, the Integrated Plant Protection Center at Oregon State University began coordinating on-farm research on beetle banks in Oregon (see www.ipmnet.org/BeetleBank/index.htm), but these have not yet been studied in the North Central United States.

Host provision

Insufficient prey or hosts will drive down natural enemy numbers. Pest populations usually rebound faster than their natural enemies, which can result in crop damage. Releasing small numbers of the pest can supply a stable food source for natural enemies without damaging the crop. One study showed that overall biological control of imported cabbageworm was more stable and provided greater benefit when adult cabbage white butterflies were regularly released. This resulted in a continuous supply of hosts for the natural enemies of this pest.

The mite complex on apples in the Midwest provides another example. Predatory mites provide biological control of pest mites such as European red mite and twospotted spider mite. Apple rust mite, which usually does not cause damage, is an alternate source of prey. If you avoid the use of pesticides that kill the apple rust mites, the predatory mites have a continuous supply of food. They will then remain in the orchard and provide continuous control when populations of the pest mites surge.

Supplying nonhost foods

Nonhost foods such as nectar, pollen, and honeydew can serve as subsistence foods during periods of host scarcity, especially for predatory insects and mites. The availability of nonhost foods may be critical to the success of many natural enemies. For example, the adults of hover flies and many parasitic insects require protein and carbohydrate foods before they can reproduce. You can provide nonhost foods either by ensuring that naturally occurring sources are continually present or by adding artificial foods to the environment.

Flowering plants can be important sources of nectar and pollen (figures 116–118), but the flower structure must be such that natural enemies can access these important foods and the plants must be in bloom when the natural enemies need supplemental foods. The total area of flowers is one of the most important factors influencing natural enemy attraction to a flowering plant or plot of flowering plants. Many species in the carrot, aster, legume, and mustard families are attractive and useful to various natural enemies (see table 8). An added benefit is that many of these plants also attract important pollinators including a wide variety of bees. Some seed companies produce seed mixtures of plants that are known to be attractive to natural enemies and that bloom throughout the growing season. Planting these mixtures in and around cropped areas can attract and retain natural enemies and pollinators. To support natural enemies in spring and early summer, it might be necessary to plant selected perennial species, since few annuals bloom early enough in the season. Michigan State University has a comprehensive website on using native perennial plants to enhance natural enemies and pollinators (www.nativeplants.msu.edu). Another useful resource is USDA publication *Farmscaping to Enhance Biological Control: Pest Management Systems Guide* (attra.ncat.org/attra-pub/PDF/farmscaping.pdf).

Figure 117. Habitat plantings provide shelter, alternate prey, and nectar and pollen for many natural enemies.

Table 8. Some flowering plants known to attract natural enemies

Common name	Scientific name	Plant type	Native?
Apiaceae (carrot family)			
ammi or bishop's flower	*Ammi majus, A. visnaga*	annual	
angelica	*Angelica atropurpurea*	perennial	+
caraway	*Carum carvi*	biennial	
coriander	*Coriandrum sativum*	annual	
cow parsnip	*Heracleum maximum*	perennial	+
fennel	*Foeniculum vulgare*	annual, tender perennial	
golden alexanders	*Zizia aurea*	perennial	+
lovage	*Levisticum officinale*	perennial	
wild carrot (Queen Anne's Lace)	*Daucus carota*	biennial	
wild parsnip	*Pastinaca sativa*	biennial	
Asteraceae (aster family)			
aster	*Aster* species	perennial	+ (some)
blanketflower	*Gallardia* species	perennial	
common boneset	*Eupatorium perfoliatum*	perennial	+
coneflower	*Echinacea* species	perennial	+
coreopsis	*Coreopsis* species	annual, perennial	+
cosmos	*Cosmos* species	annual	
cup plant	*Silphium perfoliatum*	perennial	+
goldenrod	*Solidago* species	perennial	+
sunflower	*Helianthus* species	annual, perennial	+
	Heliopsis species	perennial	+
yarrow	*Achillea* species	perennial	+
Brassicaceae (mustard family)			
basket-of-gold alyssum	*Aurinium saxatilis*	perennial	
mustards	*Brassica* species	annual	+ (some)
sweet alyssum	*Lobularia maritima*	annual	
yellow mustard	*Barbarea vulgaris*	biennial	
Fabaceae (legume family)			
alfalfa	*Medicago sativa*	perennial	
fava bean	*Vicia faba*	annual	
sweet clover	*Melilotus alba, M. officinalis*	biennial	
vetch	*Vicia grandiflora, V. villosa*	varies by location	
Other plants			
blue lobelia	*Lobelia siphilitica*	perennial	+
buckwheat	*Fagopyrum esculentum*	annual	
Canada anemone	*Anemone canadensis*	perennial	+
cinquefoil	*Potentilla* species	perennial	+
flowering spurge	*Euphorbia corollata*	perennial	+
horsemint / spotted bee balm	*Monarda punctata*	perennial	+
Indian hemp	*Apocynum cannabinum*	perennial	+
meadowsweet	*Spirea alba*	perennial	+

It is also possible to provide food for natural enemies by spraying artificial foods on plant surfaces. Central American subsistence farmers have been known to apply dilute sugar solutions to their crops to attract and retain natural enemies. Several artificial foods for natural enemies are also commercially available. The first of these, called Wheast, consists of yeast, whey, sucrose, and water and was initially developed to improve the effectiveness of green lacewings. Adult lacewings require foods such as nectar or honeydew and may not lay eggs in the absence of such foods. Wheast or similar sugar sprays can significantly increase the benefit not only of lacewings, but also of other predators, such as syrphid flies and lady beetles, and certain parasites.

Reducing interference with natural enemies

Several factors may reduce the effectiveness of or kill natural enemies. By being aware of these factors, you may be able to make management decisions that tip the balance in favor of natural enemies. The following are a few examples.

Natural enemies have their own natural enemies

Beneficial predators and parasites have their own natural enemies that can kill or interfere with the beneficial natural enemies. For example, some ants "farm" various aphids, mealybugs, and scale insects to harvest the energy-rich honeydew they produce. As they tend their "stock," the ants disrupt biological control by killing or driving off natural enemies. When you see this happening, the use of appropriate chemical insecticides to spot treat ant nests can substantially improve biological control. If the ants are tending aphids in a tree, insecticide use can be avoided by using a barrier such as Tanglefoot around the tree trunk and by pruning branches so that they do not touch the ground.

Cultural practices can hinder natural enemies

Dust can impair the activity of small predators and parasites. In orchards and vineyards, predatory mites that control spider mites are less effective when foliage is dusty. Reducing cultivation during dry periods and eliminating road dust improves biological control.

Tillage can kill or interfere with the activity of natural enemies of soil insects as well as natural enemies that overwinter in the soil. Reduced tillage systems usually support more of these natural enemies.

Host plant characteristics can hinder natural enemies

The physical and chemical characteristics of plants can affect natural enemies. For example, leaf hairs can interfere with small predators and parasites. Hooked and sticky hairs of tobacco leaves interfere with egg parasites; similar hairs on cotton leaves interfere with the egg parasite *Trichogramma* and predatory green lacewing larvae. On the other hand, predatory mites are more abundant on apple and grape cultivars that have fuzzy or hairy leaves with pronounced veins. Biological control of pest mites has proven much less successful on fruit crops whose leaves are hairless.

Insect predators have difficulty walking on cabbage cultivars whose leaves have a thick waxy coating. It is also thought that spores of insect-pathogenic fungi do not adhere well to pea leaves with a thick waxy coating. In both cases, it appears that biological control of insect pests could be improved by the planting of lower-wax cultivars. Other physical traits of plants that affect natural enemy activity include color and plant architecture (height, shape, canopy density, etc.). However, no specific resistant traits have broad application and very few plant breeders have considered the needs of natural enemies when developing new cultivars.

The chemical characteristics of plants can also be important. For example, larvae of the parasitic wasp *Cotesia congregata* can be sickened or killed by nicotine. The larvae ingest nicotine through the host insect, tobacco hornworm, which feeds on tobacco leaves. Tobacco cultivars with lower nicotine content have a higher percentage of successful hornworm parasitism by *C. congregata*.

Pesticides affect natural enemies as well as pests

Of all the areas of conservation of natural enemies, pesticide impacts have received the most study, and changing the types of pesticides used and usage patterns has been the most successful strategy. For example, some Christmas tree growers have switched from broad-spectrum insecticides to spinosyns for sawfly control, because the shorter residual and more selective spinosyns conserve predators of spider mites.

Although many people consider only the effects of broad-spectrum insecticides, other pesticides, including those allowed in organic farming systems, have impacts on natural enemies as well. For example, some fungicides used to control plant diseases can interfere with insect-pathogenic fungi. The fungus *Entomophthora muscae* gives natural control of adult flies of the onion maggot. Certain fungicides, such as chlorothalonil, copper sulfate, and maneb, reduce the effectiveness of this fungus. Some fungicides even have insecticidal or acaricidal properties and can interfere with natural enemies. The commonly used orchard fungicide thiophanate-methyl suppresses populations of predatory phytoseiid mites. Even the use of herbicides can affect natural enemies if they eliminate plants that provide necessary shelter, food, or alternate prey.

Pesticides can interfere with biological control in the following ways:

Natural enemy mortality. In general, natural enemies tend to be more susceptible to pesticides than plant-feeding insects, which may possess detoxification mechanisms to cope with defensive compounds produced by their food plants. In some cases, all stages of natural enemies are killed at the time of application, but sometimes only exposed stages are affected. Most pesticides leave an active residue for a period after application; this period can range from hours to weeks, depending on the pesticide and various conditions such as temperature, light, and precipitation. Active residues continue to kill natural enemies, including those that were in a resistant stage at the time of application and those that migrate into the sprayed area. If the pesticide does not kill the natural enemy immediately, it could accumulate to lethal levels as the natural enemy feeds on exposed prey. Parasite larvae that are inside their host insects may or may not be killed by the insecticide directly, but if their host is killed by the insecticide, they will not be able to complete their development to the adult stage.

Nonlethal effects. Pesticides may not kill a natural enemy, but may stress or weaken it or change its behavior so that it is no longer effective. Nonlethal effects include lengthened developmental time of immature natural enemies, reduced prey consumption, and reduced reproductive capability.

Indirect effects. When pesticides eliminate the hosts or prey of natural enemies, they must leave the area to find new hosts or prey. Therefore, they will not be present to control a resurging population of the target pest or secondary pests. Pesticides can also reduce populations of nonpest species that serve as alternate food sources for natural enemies, and herbicides can eliminate plants that provide pollen, nectar, and favorable environmental conditions for natural enemies. Indirect effects of pesticides can sometimes involve complex mechanisms. For example, routine fungicide applications in apple orchards prevents the growth of lichens on the trees' bark, preventing the establishment of stable communities of lichen-feeding arthropods, which in turn would have supported numerous generalist natural enemies in the trees.

Summary of effects of major insecticide groups on natural enemies

Broad-spectrum **synthetic organic insecticides** have been the most widely used in both farming and gardening since their introduction in the 1940s and 1950s. Because of their broad spectrum of insecticidal activity, these insecticides are the hardest on predatory and parasitic natural enemies. Historically, these have included organochlorines, organophosphates, carbamates, and, since the 1970s, the synthetic pyrethroids. Organochlorines have been largely discontinued because of their environmental persistence, and, since the U.S. Congress adopted the Food Quality Protection Act in 1996, the use of organophosphates has been significantly reduced. Synthetic pyrethroids are still in significant use, and comprise one of the major groups of synthetic insecticides; they are very detrimental to beneficial natural enemies.

Neonicotinoids are a relatively new class of synthetic organic insecticides but already are in common usage in both farming and gardening. Although neonicotinoids show some degree of selectivity, as a group they should still be considered moderately hazardous to natural enemies.

Additional new groups of synthetic insecticides are being developed and marketed. Many of these are considered to be reduced-risk materials because they tend to be more selective and have fewer nontarget impacts than conventional broad-spectrum materials. However, each new chemical group and each new product needs to be evaluated independently for its impacts on natural enemies. At least there is a promising tendency toward more selective materials that will better fit into integrated pest management programs.

Inorganic insecticides are a small but diverse group of compounds that work by various mechanisms. Wettable sulfur is effective against mites and thrips, hydrated lime can be used against leafhoppers, and cryolite is a stomach poison that kills a variety of chewing insects. These compounds generally do not directly kill natural enemies, but their presence on foliage may act as a physical irritant, especially to smaller natural enemies. Their residual effects on pests and natural enemies last as long as the material remains on the foliage.

The effects of **insect growth regulators** vary. Some are highly specific to a particular order of pests (such as the Lepidoptera) and these are relatively safe for natural enemies. But others can interfere with the larval development of parasites and certain major predator groups, such as green lacewings.

The impact of **microbial insecticides** varies depending on the selectivity of the insect pathogen that constitutes the active ingredient. Microbial insecticides based upon viruses and bacteria are fairly selective and usually not pathogenic to most natural enemies. Some insect-pathogenic fungi have a much broader spectrum of activity and are known to kill a variety of predatory and parasitic insects. Also keep in mind that, like chemical insecticides, microbial insecticides kill the prey of predators and the hosts of parasitic insects, requiring the natural enemies to leave the area in search of food.

Microbially derived insecticides also vary in their impact. Spinosad is an insecticide produced by culturing a naturally occurring soil organism. The commercial product includes the toxin, but not the actual microorganism. Spinosad is relatively benign to predators but can harm parasitic wasps, so its use should be restricted to situations where parasitic wasps are relatively unimportant. As we write this publication, research is underway to chemically enhance the active ingredient in spinosad, making it a more effective insecticide; the implications for natural enemies have not yet been determined. Avermectin is another microbially derived product with insecticidal and acaricidal activity; it is harmful to some important groups of natural enemies.

Botanical insecticides such as nicotine, pyrethrum, and rotenone are moderately hazardous to natural enemies, but they have short residual activity; neem is significantly less hazardous to natural enemies. Although botanical insecticides may kill many natural enemies at the time of application, those that move into the treated area shortly after application are mostly not affected. The rate of recolonization depends in part on the size of the area treated. If an entire large field is treated, successful migration will be limited and slow, and there may be a resurgence of pests. On the other hand, spot treatments of small areas are usually less troublesome because natural enemies can recolonize more rapidly from adjacent untreated areas.

Insecticidal oils applied as dormant sprays are relatively harmless to natural enemies, because few natural enemies are present at the time of application. Summer spray oils are also generally safe, but may kill certain types of natural enemies, such as predatory mites.

Figure 118. A habitat planting adjacent to a vegetable production area.

Insecticidal soaps have varying levels of impact on natural enemies, but like botanical insecticides, they do not have residual activity.

Inert dusts, such as diatomaceous earth, boric acid crystals, silica gel, and kaolin clay, have varying effects on natural enemies; their residual effects will persist for as long as the dust remains on the treated surfaces.

Some specific **miticides** are relatively safe for natural enemies, but others, such as dicofol are hazardous to predatory mites.

Reducing impacts of pesticides
The most obvious way to reduce pesticidal impacts on natural enemies is to reduce pesticide use. This can be done by using pesticides only when and where necessary. For pests that immigrate into a field or orchard from surrounding areas, perimeter sprays may be effective. A trap crop planted around the edge of a field before the main crop is planted can attract and concentrate pests, allowing a targeted pesticide application. In orchards, alternate row spraying can be effective in some situations.

Some pest insects can be attracted to traps baited with specific food baits, food odors, or dispensers of their sex pheromone. Pesticides applied with the attractant will kill the attracted pests without harming natural enemies. This strategy has been developed for ants, roaches, cucumber and corn rootworm beetles, and various fruit flies.

When selecting pesticides, choose those that have lower impacts on natural enemies. If you know the most important natural enemies in your system, it may be possible to time applications so that the target pest is in a susceptible stage but the most important natural enemies will not be vulnerable. So-called "pesticide windows" have been developed to conserve the important natural enemies of spotted tentiform leafminer on apples and major parasites of alfalfa weevil.

A few commercially available natural enemies, especially predatory phytoseiid mites, have been developed with resistance to one or more types of pesticides. These can be used successfully in integrated pest management programs that require pesticide use. For more information on availability and use of these pesticide-resistant natural enemies, contact your natural enemy supplier.

The periodic release of natural enemies

Natural enemies do not always occur in adequate numbers to provide satisfactory pest control due to natural environmental factors and/or agricultural practices. One possible control method is to increase the natural enemy population by periodically releasing additional natural enemies. These may be the same species that exist at the site or they may be different species or strains. This method of biological control is called *augmentative biological control*. Many companies produce insect pathogens and nematodes as microbial insecticides, and other companies, called insectaries, mass-produce predatory and parasitic insects and mites.

There are two broad approaches to augmentation of natural enemies. The first, **inundation**, involves releasing large numbers of natural enemies for immediate reduction of a damaging or near-damaging pest population. It is analogous to a corrective insecticide application; the expected outcome is immediate pest control. The inundative release of predatory and parasitic insects is recommended only in certain situations because of the expense and the nature of their activity. One parasite that has been effective in inundative releases is *Trichogramma* for controlling eggs of various moths.

Alternatively, **inoculation** involves releasing small numbers of natural enemies at prescribed intervals throughout the pest period, starting when the pest population is low. The natural enemies are expected to reproduce to provide more long-term (but not permanent) control. The two goals of inoculative releases are: (1) to keep the pest at low numbers, never allowing it to approach an economic injury level, and (2) to keep the natural enemy present continuously in the crop. To achieve these goals simultaneously, it is essential that enough natural enemies are released to reduce the pest population but not to eradicate it. If the pest is driven to extinction, the natural enemies will perish also, leaving the crop vulnerable to subsequent infestations. Inoculative releases are commonly used in controlling various greenhouse pests, but have also been used extensively in the field, such as using *Pediobius foveolatus*, an egg-parasite of the Mexican bean beetle.

Distinguishing between inundative or inoculative releases is helpful, since these two approaches to augmentation differ in timing and rates of natural enemy releases. Application of a microbial insecticide qualifies as augmentative biological control, since the pathogen is a living natural enemy that infects, multiplies within, and usually kills the pest. But is such an application inoculative or augmentative? For it to be considered inoculative, the pathogen must go on to infect pests beyond those present when the initial application was made. Some insect viruses and some nematode applications act this way, and, at least historically, some applications of milky disease products worked inoculatively to control Japanese beetle infestations. On the other hand, most microbial insecticides, including those containing *Bacillus thuringiensis*, rarely go on to cause subsequent infections. These would best be con-

sidered inundative releases, since the natural enemy does not provide ongoing control of the pest.

Augmentative natural enemy releases are neither low-input nor sustainable. They require a relatively high input of time, labor, and money and usually must be repeated at least annually and often several times per growing season. However, when compared to the use of broad-spectrum insecticides, augmentative biological control has certain advantages, such as reduced hazard to people, the environment, and other beneficial organisms. If you are considering purchasing and releasing natural enemies, you must evaluate both the constraints and the benefits to determine if augmentative biological control is appropriate for your needs.

Common uses of augmentation

Several companies produce and supply natural enemies for augmentative release (more information about suppliers of natural enemies is located at the end of this chapter). However, relatively few types of natural enemies are commercially available. The types of crops and pests on which commercially produced natural enemies are most frequently used can be summarized as follows.

Greenhouses

The use of natural enemies has been particularly successful for control of pests of greenhouse crops. The augmentation approach is well suited to greenhouse cultivation of crops for several reasons. First, a greenhouse is a relatively closed environment, with little movement of insects from the outside. This allows for better management of pests and natural enemies. Second, the environmental conditions in the greenhouse can be regulated not only for plant growth, but also to provide optimum conditions for natural enemy development. Third, many greenhouse pests have been heavily treated with insecticides and have developed resistance to several pesticides, so growers have incentive to explore alternate control measures. Finally, the cash value of greenhouse crops is high. A little

extra expense for pest management can be more easily absorbed into the overall production cost of such high-value crops.

Research on biological control in greenhouses started in the 1960s in Europe, where greenhouse production of food is a major form of agriculture. Today, in many European greenhouse operations, biological control through the release of natural enemies is the dominant form of pest control. In the United States, most of our fruit and vegetable production is in fields rather than greenhouses, even in the winter in the mild climates of the Southeast and Southwest. Greenhouse cultivation is used primarily for ornamentals, such as cut flowers, poinsettias, and foliage plants. Because these are grown for their aesthetic attributes, consumers do not tolerate damage, including insect injury. Growers have been reluctant to use natural enemies for control because they assume that a few pests will survive and cause some aesthetic injury. However, as new methods are developed to effectively use natural enemies, and as more greenhouse pests develop resistance to pesticides, biological control is becoming an increasingly important component of greenhouse pest management. North Central Regional publication *Biological Control of Insects and Other Pests of Greenhouse Crops* (NCR581) describes effective augmentative approaches to numerous greenhouse pests; see the appendix for a list of additional readings.

Filth flies

A second major application of augmentative biological control is unusual in that it is not targeted against plant pests. Commercial insectaries supply natural enemies for the control of filth flies and biting flies, such as the house fly and stable fly. These natural enemies are used primarily on poultry production facilities, feedlots, and dairy operations. Much research has been conducted in this area, and several species of natural enemies are commercially available. Filth fly natural enemy augmentation programs are most effective when incorporated in an overall integrated pest management system that also includes proper manure-handling practices.

Orchards and vineyards

Orchards and vineyards are high-value crops with many serious pests, but unlike greenhouses, no physical barriers keep out invading pests. Therefore, use of natural enemies is not always as effective as hoped. Notable exceptions include biological control of scale insects and mealybugs on citrus in California and elsewhere. Several species of natural enemies are commercially available for citrus crops. Mass releases of the egg parasite *Trichogramma* have been used to control codling moth on apples in Russia and Europe. Codling moth granulosis virus is commercially available in the United States, and other microbial insecticides are effective against several caterpillar pests.

Vegetables and small fruits

Many pests of vegetable and small fruit crops can be controlled by releases of generalist natural enemies such as *Trichogramma* and lacewings, in conjunction with sprays of microbial insecticides containing *Bacillus thuringiensis*. The release of predatory mites for controlling twospotted spider mite on strawberries is a common practice in California, and it is worthy of examination in other growing regions. Natural enemies of the caterpillars that attack cabbage and other crucifers are also commercially available, as are predators and parasites of Colorado potato beetle and Mexican bean beetle.

Field crops

Augmentation for control of pests of field crops has been used widely in eastern European and Asian countries, where the governments produce and release natural enemies, especially *Trichogramma*, over millions of acres of field crops. This practice has not yet been widely adopted in the United States, in part because of the relatively low value of the crop compared to the cost of purchasing and releasing natural enemies. However, methods for both ground and aerial release (figure 119) of natural enemies such as *Trichogramma* over large areas have been developed, and some suppliers have information on the techniques for these types of application. There has been much promising research on augmentative biological control of pests of corn and other field crops in this country, and practical approaches for certain pests have been developed. However, interest in biological control of pests of conventionally grown field crops has declined substantially since the introduction of transgenic crops containing the Bt toxin, such as Bt corn and Bt cotton. Augmentative biological control should continue to be of interest to growers for organic markets or for foreign markets that refuse transgenic crops. The current status of augmentative biological control against some major field crop pests is discussed in the Kansas State University publication *Biological Control of Insect Pests on Field Crops in Kansas* (www.oznet.ksu.edu/library/entml2/MF2222.pdf).

Home gardens

Home gardeners are increasingly using natural enemies to protect fruit and vegetable crops as well as landscape plants. Some natural enemy suppliers sell packages of assorted natural enemies intended specifically for the home garden or small commercial farm. Hopefully, by now, you realize that biological control is not a "one-size-fits-all" approach to pest management, and that such packages are likely to be only marginally useful, at best. The packages may not contain the right enemies for your pests, or the pests may be at a different life stage than what the natural enemies attack. A better approach is to know the important pests of your garden and to seek guidance from commercial suppliers to determine which natural enemies will be most useful, and the appropriate time(s) to make releases.

There are several other specialized targets for which natural enemies are commercially available for augmentative releases. Examples include gypsy moth parasites, natural enemies of fire ants, and predators and parasites of stored grain pests. The following section contains additional examples.

Commercially available natural enemies

Approximately 100 species of predatory and parasitic insects are commercially available in the United States. In addition, insect-parasitic nematodes and a variety of insect pathogens formulated as microbial insecticides are also available. Although this may seem like a large number, consider that there are over 700 species of serious arthropod pests (insects and mites) in North America, and most of these are attacked by multiple species of generalist and specialist natural enemies. Unfortunately, effective specialist natural enemies are not currently commercially available for most pests.

The production of natural enemies requires highly specialized equipment and methods and is relatively labor intensive. Years of university, federal, and private research are often necessary to understand the biological needs of a natural enemy sufficiently to be able to mass-produce it. To assure themselves a profit, producers primarily supply generalist natural enemies of proven abilities (such as lacewings) or specialized natural enemies for major markets (such as predatory mites and whitefly parasites). It should be realized that most pests are *not* appropriate targets for the natural enemies that are currently available. However, the list of commercially available natural enemies will increase as long as producers have acceptable markets for their new products.

The following is a discussion of some of the most important natural enemies currently available. Those that are not appropriate for use in the Upper Midwest, such as parasites of pests of citrus and cotton, are not included. Chapter 6 contains additional illustrations of many of these insects.

Figure 119. This remote-piloted aircraft releases natural enemies such as green lacewing eggs and *Trichogramma*, and can treat up to 50 acres. Full-sized aircraft can be used to cover much larger acreages.

Predatory insects and mites

Lady beetles

The most commonly available lady beetle is the convergent lady beetle, *Hippodamia convergens* (figure 120). Both the adults and the larvae are predatory; their preferred food is aphids, but they will also feed on other small, slow-moving, soft-bodied insects and mites. This species is native to much of the United States, including the Midwest. The convergent lady beetle is not commercially produced; instead, it is mass-collected by the bucketful from amazingly large hibernating clusters in the hills and mountains of central and southern California. The natural behavior of this species in California is to fly out of its normal feeding grounds in the coastal and inland valleys into higher, cooler elevations, where it mass-congregates for hibernation. After this dormant period, which can run from May or June through the following February, it returns to lower elevations, where it begins feeding on aphids and other prey.

Field-collected convergent lady beetles rapidly disperse after release and may not even resume their predatory behavior until they deplete their overwintering energy reserves. A related problem is food availability: lady beetles will disperse in search of more productive hunting grounds if aphids or other appropriate foods are not abundant. Some commercial suppliers now precondition these lady beetles by temperature treatment and feeding to try to reduce the normal dispersal behavior. Nevertheless, releases of convergent lady beetles in field settings often do not result in significant reductions in pest numbers and, therefore, are not generally recommended. On the other hand, inundative releases of field-collected convergent lady beetles in screened greenhouses may help reduce large aphid populations. If you do choose to release convergent lady beetles, be aware that releases may be more efficient when large areas are treated than when smaller areas such as home gardens are targeted. In addition, you can attempt to reduce the likelihood that they will disperse by releasing them in the evening, because they do not fly at night.

Figure 120. A half-pint package of the convergent lady beetle, *Hippodamia convergens*, contains approximately 4500 lady beetles, supposedly enough to treat 1500 square feet. The beetles disperse rapidly after release, making their benefit questionable.

The mealybug destroyer, *Cryptolaemus montrouzieri*, is much smaller than the convergent lady beetle and is more specialized, feeding primarily on mealybugs. They have been used commercially in California citrus groves for many years. They have also given good control of mealybugs on a variety of ornamental plants in commercial and hobby greenhouses and in interior plantscapes, particularly in warm, humid conditions where mealybug numbers are high. The mealybug destroyer is pictured in chapter 6 (figures 37–38).

Stethorus punctillum is a specialized mite predator, especially of European red mites on fruit trees. It is less well studied as a predator in greenhouses, but it has become established and reproduced while feeding on this pest in greenhouse cucumber and pepper crops. Adults are tiny (~1.5 mm) dark brown to black beetles that can live for up to 2 years.

Delphastus pusillus and *D. catalinae* are easily confused species that somewhat resemble *Stethorus*, but are specialized predators on whiteflies. Adults and larvae will eat spider mites and whitefly nymphs, but their preferred food is whitefly eggs. Females need to consume upwards of 200 whitefly eggs per day to support reproduction, so these *Delphastus* species are more effective natural enemies at high whitefly populations. Interestingly, both adults and larvae avoid eating whitefly nymphs that contain later stages of wasp parasites.

Other species of lady beetles are often commercially available.

Lacewings

Green lacewings are somewhat specialized for feeding on aphids, but they will also feed on an assortment of other small insects and mites. They are usually released as eggs (figure 121) or young larvae, so they do not fly away. The larval stage, called the aphidlion, is the most important stage for pest control. Although the adults of some species also feed on insects, adults of all species require other food sources such as honeydew and nectar. If these foods are not available, they will fly elsewhere to lay eggs. Two species of green lacewings are commonly available. *Chrysoperla carnea* (also called *Chrysopa carnea*) is usually best for row crops and field crops, while *C. rufilabris* is better adapted for orchards. Other species may also be available; check with suppliers for the best species for your conditions. The various stages of green lacewings are illustrated in chapter 6, figures 42–47 .

Predatory mites

Most commercially available predatory mites are members of the family Phytoseiidae and are sometimes called phytoseiid mites. Several species are used to control pest mites (figure 122). These are purchased primarily for greenhouse crops and strawberries, but they can be used wherever spider mites are a problem. They are best adapted to moderate temperatures (70–90°F) and higher humidity (>70% relative humidity). However, mites vary in their environmental tolerances, which is one reason for the availability of the different species. The most commonly used phytoseiid mite is *Phytoseiulus persimilis*, which is a highly specialized mite predator that will starve in the absence of spider mites. Other predatory mites can survive on a diet of alternate food sources, such as nectar or pollen. A mixture of species may provide better, more lasting control than a single species. Predatory mites are tiny—about 0.5 mm ($^1/_{50}$ inch) when fully grown (chapter 6, figures 102–103). Many different species of predatory mites are now commercially available; check with your supplier for the best species or combination for your circumstances.

Neoseiulus (=*Amblyseius*) *cucumeris* and *N. barkeri* (=*N. mckenziei*) are good predators of thrips, which are important and difficult-to-control pests of many greenhouse crops. These species will also feed on spider mites and other small insects.

Another predatory mite that is commercially available is *Hypoaspis miles*, in the family Laelapidae. It is a native North American soil inhabitant where it feeds on many tiny arthropods. It is an effective predator of fungus gnat eggs and larvae.

Predatory stink bugs

The spined soldier bug (chapter 6, figures 21–23), *Podisus maculiventris*, kills its prey by sucking out its body fluids. They have been used for control of Mexican bean beetle and Colorado potato beetle, as well as various caterpillar pests. *P. maculiventris* is available as eggs or nymphs.

Minute pirate bugs

Minute pirate bugs, *Orius* species (chapter 6, figures 13–14), are tiny predatory bugs that feed on mites, small insects such as aphids and thrips, and insect eggs. They can be used outdoors or in the greenhouse. These bugs occasionally bite humans, and their bite can be noticeably, though temporarily, irritating.

Xylocoris flavipes is a small predator in the minute pirate bug family. It is a sucking insect that feeds on the eggs and young larvae of various moths and beetles that infest stored grain, including the red and confused flour beetles and sawtoothed grain beetles. One limitation is that it cannot penetrate hard materials to attack weevils that infest grain and dried legumes.

Figure 121. Each cup contains approximately 1000 eggs of green lacewings. The bran filler protects the eggs during shipping and facilitates uniform application.

Figure 122. Each vial contains 100 predatory mites, *Phytoseiulus persimilis*. The bran filler protects the mites during shipping and facilitates uniform application.

Predatory flies

Aphidoletes aphidimyza is a small gall midge whose larvae (chapter 6, figure 51) are predators of aphids. They often appear in aphid infestations outdoors, but can be released with great success in greenhouses. Because they pupate in the soil, they are most effective where soils are cool, moist, and loose. *Feltiella acarisuga* is another gall midge species, similar in many ways to *A. aphidimyza*, except that it attacks spider mites.

Praying mantids

Some suppliers sell mantid egg cases. Each egg case will give rise to many young mantids. Mantids are generalist predators that make no distinction between pests or beneficial insects. Their numbers usually decline rapidly after hatching, and they generally provide little if any significant value in pest control. The use of purchased mantid egg cases is not recommended for commercial agriculture, and their value in the home garden is also questionable.

Parasitic insects

Trichogramma

Trichogramma wasps (Chalcidoidea) are the most commonly used parasitic insects in augmentation programs worldwide. Millions of acres in eastern Europe and Asia are treated annually with one or more species of this tiny wasp. All species in the genus *Trichogramma* parasitize the eggs of other insects.

Some species of *Trichogramma* are specialized parasites, whereas others are more general in their acceptance of host eggs. Eggs of moths and butterflies are most frequently attacked, although some species also attack eggs of beetles and other insects. At least six species are commercially available in the United States, primarily for use against the eggs of various moths (figure 123). *T. pretiosum* is most suited for use in field and row crops, where it attacks the eggs of pests including cabbage looper, diamondback moth, corn earworm, and European corn borer. On the other hand, *T. minutum* has been released in orchards and forests against the eggs of pests like codling moth and spruce budworm. In New York, one early season release of *T. ostriniae* in sweet corn reduced European corn borer damage by 50% at the same cost as a single insecticide application. Other species of *Trichogramma* exhibit different searching behaviors and host relationships. Check with your supplier or Extension service for the species or strains best suited for your situation.

Trichogramma releases should coincide with the earliest flights of moths of the target species. Careful monitoring of pest adults, especially through the use of pheromone traps and degree-day models, will indicate the beginning of the egg-laying period and the best time to make releases.

Figure 123. A card, measuring 25 mm (1 inch) square, holding 4000 pupae of *Trichogramma minutum*. Place 10–50 cards per acre uniformly throughout the crop. *Trichogramma* are also available loose for mechanical application, such as by aircraft.

Encarsia

Encarsia formosa (figure 124) (Chalcidoidea), is a parasite of greenhouse whitefly, *Trialeurodes vaporariorum*. It is highly effective against this pest and somewhat effective against other whitefly species, such as sweetpotato whitefly, *Bemesia tabaci*. Although used primarily in greenhouses, *Encarsia* can also be used successfully against greenhouse whitefly in outdoor plantings of vegetables and ornamentals. Greenhouse whitefly occurs in all stages simultaneously (eggs, nymphs, and adults). It is therefore best to make serial releases of the parasite, which attacks only the nymphal stages. Like *Trichogramma*, the adult *Encarsia* wasp is tiny, less than 1 mm in size. You can confirm its effectiveness by observing the color of the whitefly nymphs: healthy nymphs are translucent or pale yellow, whereas parasitized nymphs turn black (figure 82). For more details, see North Central Regional publication *Biological Control of Insects and Other Pests of Greenhouse Crops* (NCR581, learningstore.uwex.edu).

Parasites of cabbage caterpillars

Several parasites are available for controlling the various caterpillars that chew holes in cabbages and other crucifers. Among these are the egg parasites, *Trichogramma*, mentioned earlier. Others include *Cotesia marginiventris*, a parasite of cabbage looper, and *C. plutellae*, *Microplitis plutellae*, and *Diadegma insulare*, all of which attack diamondback moth larvae (the first three species are Braconidae, the last is Ichneumonidae). You can use these parasites in conjunction with sprays containing *Bacillus thuringiensis*. For more details on augmentation and other biological control approaches for various pests of brassicas, see North Central Regional publication *Biological Control of Insect Pests of Cabbage and Other Crucifers* (NCR471, learningstore.uwex.edu).

Parasites of leafminers

Liriomyza leafminers (family Agromyzidae) are important pests of chrysanthemums and other greenhouse crops as well as several vegetable crops. Two species of parasitic wasps are commercially available for control of these leafminers, and the two species differ in a few significant regards. *Dacnusa sibirica* (Braconidae) prefers cooler temperatures (65–75°F) and its adults do not host feed. In contrast, *Diglyphus isaea* (Chalcidoidea) does best in warmer temperatures (75–90°F) and kills about as many hosts by host feeding as it does by parasitism. The wasps are highly specialized parasites of *Liriomyza* leafminers and are not effective against spinach leafminer, birch leafminer, spotted tentiform leafminer, or other common leafminer pests. For more details, see North Central Regional publication *Biological Control of Insects and Other Pests of Greenhouse Crops* (NCR581, learningstore.uwex.edu).

Parasites of stored-grain pests

Anisopteromalus calandrae (Chalcidoidea) is a parasite of weevil larvae found in stored grain. *Bracon hebetor* (Braconidae) is a parasite of the larvae of several moth species, including the Indian meal moth, which is also found in stored grain. In conventional granary situations, periodic releases are necessary. The frequency of releases depends on which pests are present and the severity of the infestation. For more information on biological control in stored products, see Oklahoma State University Extension publication *Stored Product Insects and Biological Control Agents* (E-912, ipmworld.umn.edu/chapters/krischik/index.html).

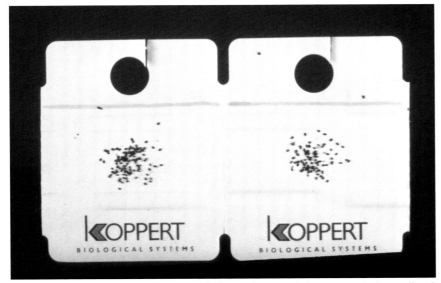

Figure 124. Approximately 150 pupae of the whitefly parasite, *Encarsia formosa,* attached to small cards that can be hung from vegetation. Place one card per 100 square foot of greenhouse or infested planting.

Parasites of filth flies

Several species of parasitic wasps are commercially available for controlling the larvae (maggots) and pupae of house flies, stable flies, and other fly species that breed in manure and garbage. These have been used most frequently around poultry farms, feedlots, and dairy farms. Such flies can be a serious nuisance or even a public-health or veterinary problem, especially where insecticide use is difficult or impossible. Commercially available parasitic wasps include *Nasonia vitripennis* (figure 77) and one or more species of *Muscidifurax* and *Spalangia* (all three are Chalcidoidea). The different species have different environmental adaptations, such as tolerance to temperature and humidity extremes. Often a combination of species is better than one alone (figure 125). The adults of these species are about 2–4 mm in size.

Because filth flies have a much shorter development time than the parasitic wasps (5–10 days as compared to 17–28 days), weekly releases of the parasites are necessary at the outset to establish a continuous presence of adult parasites. In addition, fly parasites must be used as one part of an overall integrated pest management program that also emphasizes proper manure management and control of adult flies by trapping and/or insecticide applications to the walls of buildings. (Be sure to use only legally registered products in a manner consistent with the pesticide label.) Do not apply insecticides directly to manure piles as this will kill the fly parasites.

Although several fly parasites are commercially available, not all are well adapted to conditions throughout the Midwest. For example, *Nasonia vitripennis* does not appear to be an effective parasite in Midwest feedlots and dairies. Check with suppliers to determine the best species or mix of species for your location and conditions. For more details, see Kansas State University Extension publication *Biological Fly Control for Kansas Feedlots* (www.oznet.ksu.edu/library/entml2/MF2223.pdf).

Figure 125. A shipment of 5000 parasitized fly pupae, containing mixed species of parasites.

Insect-parasitic nematodes

Nematodes are tiny worms that occur in many habitats throughout the world. Because of their small size and thin skin, they do not tolerate dry conditions. They mostly occur where there is adequate moisture, such as in bodies of water, in moist soil, and in plants and animals as parasites. Many nematodes parasitize and kill insects, and some of these have been commercialized for augmentative biological control. These are specific parasites of insects; they do not parasitize plants, humans, or other animals.

Because nematodes require significant moisture, they have been used primarily against pests inhabiting the soil, such as grubs, cutworms, sod webworms, fungus gnat larvae, and black vine weevil larvae. They have also been used against certain stem borers by injecting small quantities of nematodes into borer holes, though the labor requirements may be excessive except in small plantings.

The commercialized species are in the genera *Heterorhabditis* and *Steinernema* (=*Neoaplectana*), with *S. carpocapsae* and *S. feltiae* (*S. bibionis*) being the most widely used. Be sure to check with your supplier regarding which species is most appropriate for your situation.

In general, augmentative releases of nematodes are most effective when made during cloudy weather or in the evening, when the soil is moist and the soil temperature is 60–85°F. Watering the soil lightly after a nematode application helps wash the nematodes into the soil. Keeping the soil moist for the following week or two can also greatly increase their effectiveness.

Nematodes are commercially available in a variety of formulations, all of which are designed to be mixed and applied with water. In home gardens, they can be applied with a watering can. In commercial agriculture, they can be applied with standard pesticide application equipment, including ground sprayers, aircraft, and sprinkler systems that are approved for application of agricultural chemicals. Whatever the application method, adequate moisture must be present for the nematodes to survive and find the target pests. Nematode products have a relatively short shelf life compared with traditional pesticides; check the label for the expiration date, and carefully follow instructions for storage, mixing, and use.

Many additional species of insect-parasitic nematodes are being evaluated for commercialization, and researchers are trying to make them effective against pests in drier habitats, such as leaf-feeding caterpillars. It is likely that nematodes will have a specialized but increasing role in augmentative biological control. For more details, see The Ohio State University site, Insect Parasitic Nematodes (www.oardc.ohio-state.edu/nematodes).

Insect pathogens and microbial control

Important insect pathogens occur in several groups, including the viruses, bacteria, fungi, and protists. General aspects of these natural enemies are covered in chapter 6. As early as the 1830s scientists suggested controlling insects by distributing spores of fungal pathogens. In the late 1800s and early 1900s, farmers often collected diseased insects, ground them up in water, and sprayed them on crops that contained healthy insect pests. These early attempts were sometimes successful, but they often failed because of poor understanding of the disease cycle and the complex interrelations among pest, pathogen, and the environment. Many of these problems have been resolved with modern commercial products.

Commercial formulations of insect pathogens that are applied to crops in an augmentative manner are called microbial insecticides. Insect pathogens must have certain characteristics to be used successfully and reliably as microbial insecticides. Obviously and most importantly, they must be effective at controlling the target pest(s). Second, there must be a mechanism to mass-produce them economically, and there must be an appreciable market for the commercial product. Third, they must be easily applied and remain effective for a critical period after application. Finally, they must pass scrutiny for hazard to humans and the environment in order to receive approval by the Environmental Protection Agency. Although there are hundreds of different insect pathogens, very few have satisfied all of these criteria and been commercialized.

Bacteria
Bacillus thuringiensis

Bacillus thuringiensis (Bt) has long been known to farmers and gardeners as a control for many moth and butterfly caterpillars. In recent years, additional subspecies have been formulated into commercial products with activity against other pests such as mosquito larvae and certain beetles. The biological characteristics of Bt are summarized in chapter 6. All Bt products must be eaten by the insect in order to cause death of the host insect—they do not work by contact or fumigant action—so thorough coverage of the area where the insects are feeding is essential. Bt relies on an infection process to kill the target insect. Death does not occur immediately at the time of contact, as with many chemical insecticides. However, soon after a larva ingests Bt, its digestive system becomes paralyzed and it stops feeding. Although insects in the treated area may appear alive and healthy immediately after an application, they have stopped causing damage, and they should be dead or noticeably diseased within 48 hours of treatment (see figure 110).

The different subspecies of Bt differ in their insecticidal properties and are effective against different groups of pests. More types are discovered in nature every year. Additional subspecies will probably be commercialized in the future. The subspecies (ssp.) that are currently available commercially are summarized below.

Bacillus thuringiensis ssp. *kurstaki* (Btk) is produced and marketed under several brand names for caterpillar control. This is the most commonly used type of Bt and is available as spray concentrates, dusts, and granules for different types of applications. It is effective against many species of caterpillars on fruits, vegetables, flowers, trees and shrubs, and turf grass, and can be used to control Indian meal moth in stored grain. Younger larvae are much more susceptible than older larvae. Formulations break down and become inactive within a couple days after application. Therefore, applications must be timed properly, and additional applications may be necessary if infestations continue for an extended period. Note that although many sawfly larvae (Hymenoptera) look like caterpillars, they are not controlled by Bt.

Table 9. Some commercially available insect-pathogenic bacteria.

Species	Typical hosts
Bacillus sphaericus	larvae of mosquitoes, especially *Culex* species
B. thuringiensis ssp. *aizawai* (Bta)	greater wax moth larvae in bee hives
B. thuringiensis ssp. *israelensis* (Bti)	larvae of mosquitoes and black flies
B. thuringiensis ssp. *kurstaki* (Btk)	caterpillars
B. thuringiensis ssp. *tenebrionis* (Btt) (also *Bt morrisoni* or *Bt san diego*)	larvae of Colorado potato beetle and elm leaf beetle
Paenibacillus popillae	Japanese beetle grubs and some other white grubs; recent studies raise questions about its current effectiveness

Bacillus thuringiensis ssp. *aizawai* is similar to Btk and is even more active against some caterpillars. One of its registrations is to control the larvae of wax moths that infest honey bee hives, but it is effective against many other caterpillars as well.

Bacillus thuringiensis ssp. *tenebrionis* (Btt) is active against certain beetle larvae; in some cases it even controls adults. Commercial products containing Btt are used primarily against Colorado potato beetle larvae, which have developed resistance to conventional insecticides in some areas. Btt products are also registered to control elm leaf beetle. Although not active against all plant-feeding beetles, additional pests will probably be added to Btt labels in the future. You may also see this subspecies listed as Bt ssp. *san diego* or Bt ssp. *morrisoni*.

Bacillus thuringiensis ssp. *israelensis* (Bti) is active against the larvae of certain flies, gnats, and mosquitoes (order Diptera). It primarily controls biting flies, such as certain mosquitoes and black flies, and fungus gnats that often breed in potting soil. Control of fungus gnat larvae using Bti is accomplished primarily by applying soil drenches to infested pots. At least three weekly applications are recommended for control. Several soil factors can influence effectiveness, such as extreme levels of soluble salts, pH, moisture, and temperature.

Different formulations of Bti are available for controlling larvae of mosquitoes and black flies in aquatic habitats. Granular formulations will penetrate the canopy of trees to fall into standing water beneath, whereas liquid formulations can be applied to unobstructed areas of open water. Bti is not effective against all species of mosquito larvae, nor is it very effective in certain types of water, such as those with high levels of organic pollutants. Because mosquito and black fly adults can fly considerable distances from their aquatic breeding sites, larvicides such as Bti are most effective when treatments are done on an area-wide basis, such as by mosquito abatement districts or health departments.

Bacillus sphaericus

Like Bti, *Bacillus sphaericus* is effective against mosquito larvae. However, *B. sphaericus* remains effective in water with high levels of organic pollutants and infects some mosquito species that Bti cannot. For example, *B. sphaericus* has been used to control *Culex* mosquitoes, the primary vector of West Nile virus.

Milky disease of Japanese beetle

One of the first insect targets of microbial pesticides was the Japanese beetle, *Popillia japonica*. The larvae are white grubs that cause serious damage to the roots of turf and other grasses, and the adult beetles are major foliage and fruit pests of numerous broad-leafed plants. Early research on Japanese beetle revealed a disease called milky disease, so named because the infection causes the blood of the grub to turn milky white. Milky disease is caused by the bacteria *Paenibacillus popillae* and *P. lentimorbus*. These bacteria (formerly in the genus *Bacillus*) can be mass-produced in insect larvae and then purified and formulated into microbial insecticides; currently, only *P. popillae* is commercially available in the United States.

From 1939 to 1953, spore powders of the milky disease bacteria were used extensively to control Japanese beetle grubs in turf grass in the eastern United States. These were primarily inoculative releases, as establishment of milky disease bacteria in the soil resulted in very effective and long-lasting control. Beginning in the 1960s, however, Japanese beetle infestations again began to increase in severity, and recent evaluations of commercial milky disease products have revealed low infection rates in lab studies and essentially no control in field studies. It is unclear whether this is due to quality control issues during production of the bacteria, reduction in virulence of the bacteria, development of resistance to the pathogen by Japanese beetle populations, changes in environmental conditions, or something else. Despite past successes, the use of milky disease products as microbial pesticides should be approached with some skepticism until the reasons for reduced efficacy are resolved.

Points to remember when using *Bacillus thuringiensis*

- Correctly identify your pest. Caterpillar-active varieties do not control sawfly larvae.

- Choose the correct variety of Bt for the pest to be controlled.

- Older larvae are more difficult to control than young ones. Monitor plants frequently and make applications when larvae are young.

- Bt is inactivated by sunlight and has a short activity period. Multiple applications may be necessary for a continuing pest problem.

- Carefully read the label and follow all directions regarding storage, mixing, application, and safety. Remember that the pesticide label reflects the law; uses and practices not authorized on the label are illegal.

Various white grub species can contract milky disease but each strain or species of bacterium infects just one or a few species. Annual white grubs called chafers (*Cyclocephala* species) cause most grub damage to turfgrass in central and southern parts of the region, and these, historically at least, were somewhat susceptible to commercially available milky disease products. In the northern states, the dominant white grub species are May and June beetles (*Phyllophaga* species), which are unaffected by milky disease products for Japanese beetle. Researchers have identified other species of milky disease bacteria that attack other white grub species, as well as a new strain of Bt, *B. thuringiensis japonensis*, which is highly effective against Japanese beetle larvae, but none of these have yet been commercialized in the United States.

Viruses

Insect viruses can be highly effective when developed as microbial insecticides. They are usually more specific than other pathogens that have been commercialized, such as *Bacillus thuringiensis*. Their specificity makes them ideal for conserving natural enemies in IPM programs, but they may not be economical if multiple pests must be controlled simultaneously. As with Bt, ultraviolet light rapidly breaks down viruses. However, certain chemicals that block UV light can be added to virus formulations to improve the activity of viruses. Genetic manipulation of viruses may lead to future strains that are more persistent and faster-acting than those available today.

Several insect viruses have been developed as microbial insecticides. The United States Forest Service has produced some viruses for aerial application on large tracts of forest to control forest pests such as gypsy moth, European pine sawfly, and spruce budworm; these viruses are not available commercially. A virus of cotton bollworm was registered at one time but was not competitive with newer broad-spectrum insecticides. Commercially available insect viruses include granulosis viruses (GVs) for codling moth and Indian meal moth and nuclear polyhedrosis viruses (NPVs) for alfalfa looper, beet armyworm, Douglas fir tussock moth, gypsy moth, tobacco budworm, and pine sawflies.

Even if a specific insect virus is not commercially available, it is sometimes possible to take advantage of naturally occurring viral outbreaks. Farmers in Brazil collect virus-killed insects and store them in a freezer for future use. To treat 1 acre, 10 ml (about 2 teaspoons) of chopped-up caterpillars are blended with water and diluted to 25 gallons. When used against young caterpillars attacking cassava and rubber trees, this approach caused 90% mortality of the pests in just 4 days. Similar results should be possible in any crop as long as the pest being targeted by the homemade virus preparation is the same as the one from which the virus was obtained.

Fungi

Several insect-pathogenic fungi have been developed as microbial insecticides. (See table 10 for some commonly available products.) Because fungi need high humidity for spore germination, their use is restricted to moist habitats or situations in which the humidity is naturally high or can be artificially increased. Unfortunately, fungal plant pathogens may also thrive in high humidity environments, so you must weigh the trade-offs before deciding to increase the humidity in your cropping system.

Survival of fungal spores after application is essential for successful use of fungi as microbial insecticides. Some commercial products are formulated with oils to coat and protect fungal spores for use in low humidity environments or ultra-low volume applications. Fungi, like most other insect pathogens, can be damaged by ultraviolet radiation present in sunlight, so commercial formulations may include some degree of UV protection. Nevertheless, applications of fungal insecticides are bound to be most effective at moderate temperatures, high humidities, and on cloudy days.

Table 10. Some commercially available insect-pathogenic fungi

Species	Typical hosts
Beauveria bassiana	wide host range, including aphids, beetles, caterpillars, grasshoppers, thrips, whiteflies, tarnished plant bugs
Metarhizium anisopliae	beetles, cockroaches, grasshoppers, spittlebugs, termites
Paecilomyces fumosoroseus	wide host range, including aphids, beetles, caterpillars, grasshoppers, spider mites, thrips, true bugs, whiteflies
Verticillium lecanii (not yet available in the United States)	wide host range, including aphids, whiteflies, thrips

Innovative delivery methods have been developed for some insect-pathogenic fungi. For example, cockroach traps lined with spores of *Metarhizium anisopliae* not only protect the spores from adverse conditions, but also ensure close contact between the spores and cockroaches that visit the traps. In greenhouses, bumble bees have been used to deliver spores of *Beauveria bassiana*, resulting in significantly greater infection and mortality of tarnished plant bugs and western flower thrips.

Protists

Although some protists rapidly kill their hosts, others cause prolonged, chronic infections. These infections may ultimately result in death, but some insects may reach adulthood and even reproduce, although the number of offspring may be limited. Pathogens causing chronic infections are not useful for the immediate population control you might expect from many other microbial insecticides. For example, *Nosema locustae*, which is commercially available in a bait formulation for grasshoppers, results in a long-term infection. It can be very effective when applied over large areas, such as pastures or rangeland, especially if applied when the grasshoppers are young. The product is also sold for homeowner use, but it is not very effective at protecting garden plantings because of its slow action and because grasshoppers can rapidly invade the garden from adjacent untreated areas.

Lagenidium giganteum (formerly classified as a fungus) is commercially available for use in control of mosquito larvae, especially those in the genera *Aedes* and *Culex*. Organic pollutants do not appear to reduce the activity of this insect pathogen.

No other protists are commercially available at this time, but research continues on other species of *Nosema*, *Vairimorpha*, and several other genera.

Assessing the quality of purchased natural enemies

The mass production of any living organism can be a challenging task. For almost every species of predatory or parasitic insect raised, the producer must also mass-propagate the prey or host insect, which may also mean growing the host plants necessary for both. Occasionally, though, even the most well-conceived protocols produce less-than-desired results. Commercial insectaries must operate at a profit, and over-production results in lost profit. Furthermore, although the producer attempts to ensure a high-quality product, problems may arise during the shipping process, such as unexpected delays or exposure of the natural enemies to extreme temperatures. For these and other reasons, commercial natural enemy shipments are sometimes of less-than-optimum quality, leading to poor control.

Detailed quality-control guidelines have been established for producers of natural enemies (for example, the International Organization for Biological Control, www.amrqc.org), and most suppliers conscientiously evaluate the quality of their natural enemies for the characteristics discussed below. After all, it is in their own best interest to ensure that the natural enemies they sell are effective. However, if you purchase natural enemies for release, you also should evaluate the following characteristics, as appropriate.

Survival

Natural enemies have to be alive to be effective. In some cases, as with adult lady beetles, their general condition is easily observed. In other cases, it may be more difficult. Parasites such as *Trichogramma* and the whitefly parasite *Encarsia* are shipped in an immature stage within the bodies of their hosts, and it is very difficult to evaluate survival until the adult wasps emerge. Similarly, lacewings and some other natural enemies may arrive as eggs, and survival can only be evaluated by the percentage of eggs that hatch.

What to know when ordering natural enemies

There's a lot to know, but you can get help from Extension publications, your local Extension agent, private pest management consultants, and your natural enemy supplier.

- Know the specific pests you need to control—correct identification is crucial!

- Know the best natural enemies, either singly or in combination, for the target pest or pests.

- Know the proper timing of release of the natural enemy, based on the life cycles of both the pest and natural enemies.

- Know the proper release rate for each natural enemy.

- Calculate the amount of natural enemies needed, based on release rate, area to be covered, and severity of the pest infestation.

- Know the recommended frequency of release if multiple releases are necessary.

- Work with your supplier on shipment details to ensure that you'll be able to care for the shipment as soon as it arrives.

- Understand proper release practices so that you will be prepared to make releases when the shipment arrives.

- Understand proper storage requirements if releases cannot be made immediately after arrival.

Vigor

It is not sufficient that the natural enemies be merely alive. They have to be fully capable of undertaking their normal behaviors and life processes once they are released. Lacewing larvae, for example, need to be able to crawl considerable distances in search of their food, and parasitic wasps must be able to feed, mate, and seek out their host insects in order to reproduce and provide control. Any serious stress on natural enemies is likely to interfere with their ability to find and kill pests. If natural enemies arrive in an active state, observe them awhile to evaluate their ability to fly or crawl. If they arrive as eggs, pupae, or parasitized hosts, consider setting aside a small portion of the shipment in a sealed container to evaluate survival and vigor.

Quantity

The degree of control achieved is based on recommended release rates. Any reduction from these rates will reduce overall control. Suppliers of natural enemies often ship more than you ordered to compensate for losses during shipment and handling. In spite of the best efforts of producers, however, the number of live natural enemies emerging from a shipment may be substantially different from that ordered.

Sex ratio

Most insects have roughly equal proportions of males and females. Occasionally during mass-production sex ratios can become significantly lopsided. In the case of parasitic wasps, the adult female lays eggs in the host, ultimately killing the host. If the shipment is predominately males, host mortality may be less than desired. It is very difficult for the purchaser to determine the gender of natural enemies. We must rely on the producers to provide shipments of appropriate sex ratio.

Proper species

Many related natural enemies look similar to each other, especially the tiny ones such as predatory mites and the different species of *Trichogramma*. Although they look similar, related species may actually have quite different biological characteristics, including the species of pests they attack. As in the case of sex ratios, it is difficult for the user to distinguish among related species. Again, we must rely on the integrity and the quality-control standards of the producer to guarantee shipments of the proper species.

General considerations

Other factors influence natural enemy effectiveness, including reproductive capacity, life span, and ability to locate prey or hosts. Again, we have to trust the producer to ensure that natural enemies are high quality. If a natural enemy release has not been as effective as expected, one or more of the quality-control issues discussed above could be the cause. Consider the recommendations in the sidebar when ordering and receiving natural enemies.

Reputable producers and suppliers of natural enemies will do everything they can to help you. By helping you to use biological control successfully, they increase the chances that you will be a repeat customer. They have lots of information about which natural enemy to use and when and how to apply it. They are concerned about quality-control issues, and they are always working to improve their products. If you have questions or concerns, contact your distributor; he or she should be willing to work with you to assure that the product meets your needs. If your supplier does not resolve your problems satisfactorily, you should change suppliers.

What to do when you receive a shipment of natural enemies

- Minimize exposure to hot or cold temperatures.
- Carefully inspect the shipping container before opening, and note damage that likely occurred during shipment.
- Carefully inspect contents for damage.
- Determine if you received the species and quantities you ordered.
- Attempt to assess the quality of the natural enemies.
- Contact your supplier immediately if you are aware of any problems.
- Make releases as soon as possible after receipt.
- Store under proper conditions if it is necessary to delay releases.
- Make releases based on methods recommended by your supplier; releases should usually be made during a cooler part of the day and when rain is not imminent.
- During release, check once again for quality characteristics.
- Attempt to evaluate success of your release by monitoring natural enemy populations and impacts on pest numbers.

Cost analysis of natural enemy releases

Some natural enemies are much easier and less expensive to raise than others. This is reflected in their prices. Because of differences in prices and use patterns, it is difficult to generalize about the cost of using natural enemies. Many commercially available natural enemies have not been adequately researched for suppliers to recommend specific release rates based on pest population levels. Commercial suppliers base their guidelines as much on experience as on research. Release rates often vary depending on the size of the pest population and the developmental stage of the crop. Also, release rates for preventive inoculation programs can be substantially lower than corrective inundative releases.

The cost of augmentative releases of natural enemies may be competitive with the use of insecticides or other pest management practices. In many situations, however, the use of natural enemies will be more expensive, and you have to decide if the additional cost is justified. Keep in mind that the high value of many specialty crops reflects high production costs, including pest management. The expense of biological control for these crops may be relatively low compared to other pest management practices and overall production costs. On low-value crops, however, natural enemy releases must be inexpensive to be practical.

When considering the cost of a pest management program that incorporates augmentative releases, don't forget to include related costs such as labor for pest monitoring, sanitation, and the use of selective insecticides. Also, keep in mind that all natural enemies are somewhat specialized. If your crop routinely has serious problems with multiple pest species, then multiple natural enemies or additional pest management practices may be necessary. These all add to the overall cost of a pest management program.

To assess the costs and feasibility of using natural enemies, you should do the following:

- Calculate current pest control costs, including not just materials but also hidden costs such as labor and tractor fuel and maintenance.

- Calculate the costs of switching to a natural enemy program. To do this, contact several suppliers of natural enemies, tell them your specific requirements, and ask for specific recommendations for release rates, frequency of releases, and costs of natural enemies. If you receive different answers from different companies, ask for clarification. Natural enemy suppliers who are truly interested in biological control and the success of their customers will be willing to help you develop a program.

- Compare your current costs to those of switching to a natural enemy program. If costs of switching are higher, decide if these are justified. If you are a commercial grower, determine if you can compensate for the additional costs by adopting new marketing strategies, such as targeting the organic food market.

- When starting an augmentation program, start with a limited acreage for one or two years. After you become familiar with the system, increase your acreage at a comfortable rate. If problems arise during this period, contact your supplier, consultant, or county Extension agent for additional assistance.

- Consider hiring a private pest management consultant who is familiar with natural enemies and their use in augmentation programs. A consultant can help fine-tune your natural enemy augmentation program, and make recommendations as to appropriate natural enemies, release rates, and frequencies, based on regular crop monitoring. Unfortunately, many areas of the Midwest have few such consultants.

Approaches to developing augmentation programs

Several approaches to augmentation have been developed, depending on the types of pests and the interests of individuals, organizations, or agencies involved.

Buying directly from a supplier

The most common approach to augmentation is for the private individual—whether farmer, horticulturist, rancher, or home gardener—to buy natural enemies directly from a producer or distributor. (A distributor buys in bulk quantities from an insectary at wholesale prices and then sells in smaller retail quantities, or arranges drop-shipments from one or more producers.) To gain maximum effectiveness from this approach, the user must (1) become knowledgeable about the pest complex, including the life cycles of the pests, (2) become knowledgeable about natural enemies, both native and commercially available, (3) be willing to monitor pests frequently, (4) have a good working relationship with the natural enemy supplier, and (5) plan the year's augmentation program and place natural enemy orders well in advance. Remember, while the use of insecticides is easy and the results are rapid, this is not the case with biological control. This approach requires considerable research and advance preparation. It is not a last-minute solution.

Contracting with a pest management consultant

More and more farmers are working with pest management consultants or with general crop consultants who are competent in pest management. Unfortunately, in most parts of the country today, few crop consultants are sufficiently knowledgeable about natural enemies to make decisions based on their occurrence or to recommend natural enemy augmentation programs. This situation will improve as interest increases in more environmentally sound methods of pest control. In the meantime, organic certification agencies may be able to suggest a consultant with expertise in biological control, even if you are not an organic grower. Some insectaries and natural enemy distributors can provide pest management consultants for local growers. You may wish to check on the availability of such programs in your area.

There are many benefits to working with a private pest management consultant. Most farmers have found that they more than pay for themselves in cost savings for pest control.

Grower-cooperatives producing natural enemies

In Europe, some large greenhouse growers produce their own natural enemies. Natural enemy production in Cuba is almost a cottage industry, with small-scale producers scattered throughout the country. And for many years in southern California, three local cooperatives used to produce natural enemies for their member growers. One of these, Associates Insectary, is the only remaining cooperative insectary in the United States. The cooperative's pest control advisors inspect over 10,000 acres of citrus and avocado orchards owned by its member-growers, and recommend pest control practices. The insectary produces four natural enemies and releases them as needed in the members' orchards. Members are assessed a fee based on acreage or natural enemy needs. This type of cooperative could greatly reduce the cost and enhance adoption of biological control, especially for high-value crops that are produced extensively in a localized area.

Membership in a cooperative is voluntary. A similar but mandatory approach would be the creation of a special taxation district. Users would be taxed for the production (or purchase) and release of natural enemies throughout the district. To our knowledge, no such taxation districts exist for agricultural pests, but many have been established around the country for mosquito control, although biological control is only one approach used.

Augmentation by governmental agencies

Local, state, and federal agencies often produce or purchase natural enemies for augmentation programs. Although policy makers usually make these decisions, input from the general public can be important. State and federal agencies employ professional entomologists to aid in the decision-making process; these augmentation programs are based on the best scientific information available. This is usually not the case for local jurisdictions, and local agencies sometimes make serious blunders when using natural enemy releases. For example, some communities have attempted to control mosquitoes by releasing lady beetles, although lady beetles do not feed on mosquitoes. Local agencies should seek guidance from state or university entomologists when considering an augmentation program.

The future of augmentation in the United States

The use of biological control by augmentation of natural enemies has become an accepted and established practice. However, relatively few species of natural enemies are commercially available in the United States, and these attack a relatively small number of pest species. Natural enemy augmentation programs have had their greatest impact in western Europe in greenhouse crops, and in eastern Europe and China where *Trichogramma* is used for control of various caterpillar species on millions of acres of field crops, row crops, orchards, and forests. In the United States, the use of microbial insecticides has also become an accepted practice.

Greenhouse production is really the only crop system in which most insect and mite pests can be controlled by natural enemy augmentation. Augmentation may also be appropriate in situations where there are few if any additional pests beyond the target pest, such as with filth flies or pests of stored grains, so that natural enemy disruption by other pest management practices can be avoided. In other situations, however, as with many row crops and field crops, serious pest species exist for which no commercially available natural enemies exist. It is then up to the pest manager (that is, farmer, gardener, livestock manager, or pest management consultant) to determine if the augmentation approach for only one pest of the crop can realistically and economically be incorporated into the overall pest management for that crop. Fortunately, some of the modern insecticides have a limited spectrum of activity and are fairly safe for natural enemies, allowing for the integration of chemical controls with augmentation biological control.

In the future, we expect to see more private companies commercializing biological control, from insectary production of natural enemies to commercialization of insect pathogens to providing pest management consultants in the field. We also see great potential for the development of grower cooperatives for mass-producing and releasing natural enemies for members, but this would require a major shift in mindset. Additional natural enemy species will be added to the list of those currently available for augmentative releases. However, such advances will come slowly. This will require scientific research, practical observations of the natural enemy's pest control potential, consideration of how to incorporate the natural enemy into existing pest management programs, and how to mass-produce them cost effectively. The number of commercially available natural enemies is increasing, and this trend will continue as long as farmers and gardeners seek alternatives to chemical controls. The methods for maximizing the effectiveness of augmentative releases will improve with continued research and practical experience. Be sure to keep in touch with your natural enemy supplier, your trade organizations, and your Extension service for new advances in augmentation technology.

Suppliers of natural enemies

Predatory and parasitic insects and nematodes are available by mail from many suppliers. Because products, sources, and prices are subject to change, we recommend that you consult an up-to-date information source, such as one of the following:

- **The IPM Practitioner.** A newsletter published 10 times per year by Bio-Integral Resource Center, P.O. Box 7414, Berkeley, CA 94707. They publish an annual directory of least-toxic pest control products that lists almost all beneficial insects, mites, microbes and other pest control products that are available in the U.S., along with contact information for each supplier. Subscription and other information available at www.birc.org.

- **Association of Natural Biocontrol Producers.** Formed in 1990 to represent the interests of producers, distributors, and users of natural enemies in augmentative biological control. Their website includes some background information on biological control and a list of its members with contact information and links to websites. More information at www.anbp.org.

Microbial insecticides are available from many garden centers and agricultural chemical suppliers.

Putting it all together

Biological control can be conducted in virtually all crops, managed forests, recreational areas, natural habitats, landscape plantings, and home gardens. Although some pests cannot be completely controlled by natural enemies, existing technology allows for a much greater use of biological control than is currently practiced. New methods will be developed in response to increasing interest.

You cannot achieve successful biological control by following recipes such as those for using chemical insecticides. Many variables concerning the crop, the pest complex, and the environment will influence the effectiveness of natural enemies. Biological control is not always the perfect solution; it has its strengths and weaknesses, as do all approaches to pest control. That is why biological control must be used in conjunction with other control methods in an integrated pest management program. If natural enemies do not adequately control a pest, other approaches such as modifying a cultural practice or planting a resistant crop variety may be needed. This reduces the need for chemical insecticides that would be harmful to natural enemies controlling other pests in the crop.

How does biological control fit into an overall pest management program?

Compared to the United States, much of Europe has seen more extensive adoption of biological control and organic farming practices. Based on these experiences, some European researchers have developed the following hierarchy of tactics for pest management in organic farming systems. This hierarchical approach is relevant to any type of farming system and reminds us that biological control practices will be most successful when appropriate management practices are already reducing pest populations and creating a favorable environment for natural enemies.

Phase 1: Suppress/prevent pest problems using crop management practices that are compatible with natural processes. Such practices include selection of planting date, resistant plant cultivars, crop rotation, nutrient management, tillage, harvest practices, and location of farms and fields.

Phase 2: Conserve and enhance natural enemy impacts through vegetation management. Practices include cover crop management, crop diversification, insectary plantings, and management of plant communities around fields.

Phase 3: Use augmentative biological control, selecting inoculative and/or inundative approaches as appropriate.

Phase 4: Apply biological or mineral insecticides and use mating disruption, if available.

And to this, we would add

Phase 5: If and when it is deemed necessary to use chemical controls, use the "softest" pesticide available and apply it to the smallest area possible (e.g., spot treatments or border applications) to minimize negative impacts on natural enemies.

(Adapted from G. Zehnder, G. M. Gurr, S. Kühne, M. R. Wade, S. D. Wratten, and E. Wyss. 2007. Arthropod Pest Management in Organic Crops. *Annu. Rev. Entomol.* 52: 57–80.)

Factors to consider when planning a biological control program

Biological control will work best when you are familiar with the pests and their natural enemies. Therefore, the most important aspect of planning a new pest management program is to gather the information necessary for making the correct decisions. If you cannot make this commitment, you are much less likely to be successful. Information on natural enemies and biological control is available from reading materials, crop consultants, suppliers of natural enemies, and the Cooperative Extension Service. The internet is a valuable resource, but be certain that the sites you use are reliable.

Consider the following factors when planning your biological control program.

Effectiveness of the natural enemies

Many pests have very effective natural enemies and can readily be managed by biological control. If a pest does not have sufficiently effective natural enemies, other approaches may be necessary.

Remember that there may be a time lag before natural enemies suppress an increasing pest population. Sometimes the pest can develop to damaging levels during this period, and other actions may be needed temporarily. The best time to manipulate or augment natural enemy populations may be long before the pest population has a chance to build up.

Some natural enemies do not kill the pest until late in its life cycle, allowing it to cause some injury. Such natural enemies reduce the overall pest population by reducing reproduction, but they do not eliminate damage caused by the current generation of pests. Carefully monitor your crop to determine the effectiveness of natural enemies and whether other controls may be necessary; we discuss monitoring natural enemy effectiveness later in this chapter.

Natural enemies require the presence of a few pests for survival and reproduction. You should know what pest population levels are likely to cause economic damage and accept that some number of pests below this level is actually beneficial because they will support a continuous population of natural enemies. A few types of pests, such as those that transmit serious plant pathogens, can cause economic damage at very low population levels; use of natural enemies may not be an appropriate method to control these pests.

Sustainability

Some approaches to biological control are the most self-sustaining of all pest management practices. Natural enemies, like all living organisms, reproduce themselves, resulting in continuous pest control. However, you must assure that natural enemies have necessary resources and are not hindered from reproducing. Review chapter 8 for more information.

Safety

Biological control is one of the safest forms of pest management. Natural enemies are not toxic, pathogenic, or injurious to humans and wildlife. In general, biological controls are nonpolluting and are not as damaging to the environment as more disruptive techniques such as plowing or the use of broad-spectrum insecticides. Natural enemies leave no residues on food or in living, work, or recreational areas.

Only a few important natural enemies are capable of biting or stinging. The rare cases of injury usually result from improper handling.

Many natural enemies are highly specific to the target pest or a very limited range of prey or hosts. They will not adversely affect nontarget organisms. When new natural enemies are considered for importation and release in this country, they are carefully scrutinized by state and federal agencies to ensure that they will not themselves become pests or harm the environment. Some natural enemies do have a broad range of hosts and may attack nontarget organisms. For example, microbial insecticides containing *Bacillus thuringiensis* for caterpillar control should not be used where they will harm threatened or endangered species of butterflies and moths.

Cost

Some forms of biological control are cost-free or involve costs that are competitive with other control approaches. Establishing non-native natural enemies is funded by state and federal agencies, with little direct cost to the farmer or gardener. Establishing perennial plants to provide habitat for natural enemies involves a one-time cost with pest-reduction benefits that should last for the life of the planting. Biological control programs may involve extensive monitoring of pests and natural enemies, but these costs apply to all integrated pest management programs.

The cost of natural enemies used in augmentation programs varies with the type of natural enemy. Some may be competitive with other approaches, but others are relatively expensive and their use can be justified only on high-value crops. Crops that have multiple key pests may require several natural enemies that collectively cost more than a single broad-spectrum insecticide. In some situations, however, the other benefits of biological control justify the higher cost of releasing natural enemies.

Environmental constraints to success

Natural enemies may be adversely affected by numerous environmental factors, such as weather conditions. These factors vary by location and change through time, sometimes causing pest populations to fluctuate. Consequently, the results of biological control are sometimes less predictable than other approaches. Routine pest monitoring should give you adequate time to respond to a pest population that is approaching damaging levels.

Besides natural constraints to biological control, natural enemies may be harmed by cultural practices such as tillage, planting large monocultures, and simplification of the overall farm habitat. Biological control will not be fully effective if these practices are essential. However, detrimental practices can often be eliminated or modified to protect and encourage natural enemies.

Regional differences

By now you've probably learned that successful biological control requires the right natural enemies to be present in the right numbers at the right time relative to the development of both the pest population and the crop. This balance depends critically on environmental conditions, which can vary from place to place and from year to year. Some natural enemies may be able to complete more generations at lower latitudes or during warmer years, and this may allow larger populations to develop. Overwintering survival can be affected by extreme temperatures, amount of snow cover, or unusual midwinter thaws, and the number of insects that survive the winter has a strong influence on population growth the following season. Especially in more northern areas of the central United States, winter weather can cause local extinctions of some natural enemies, and these must re-colonize the region from milder regions each spring.

Another complicating factor is the spread of intentionally or accidentally introduced species. For example, the tarnished plant bug parasite, *Peristenus digoneutis,* was introduced into New Jersey in 1984 and has slowly dispersed to six surrounding states, but has not moved south of latitude 41°N. As this species moves into a new area, tarnished plant bug populations decline. Beginning in 1988, the non-native multicolored Asian lady beetle, *Harmonia axyridis,* spread rapidly from Louisiana, where it was first detected, to much of North America, causing reductions in populations of several native lady beetles.

Populations of insect pests and natural enemies can also be affected by the invasion and subsequent control of non-native plant species. For example, Queen Anne's lace (*Daucus carota*) was listed as a noxious weed in Ontario. Locations with successful eradication of this plant experienced more serious pine shoot moth infestations, since one of its important parasites had been using Queen Anne's lace flowers as a nectar source.

These types of variations in climate, plant communities, and insect communities make it nearly impossible to predict how the pest–natural enemy balance will tip in a given location in a given year; the one possible exception may be in greenhouses, where environmental conditions are more controlled. Biological control practitioners must be aware that information that is accurate for one location may not be entirely accurate for their locale. Regional and seasonal differences mean you need to be willing to regularly monitor the success of your efforts and to make any necessary changes.

Sampling can give you incomplete information

- Sampling during daylight hours will miss nocturnal natural enemies.

- Sweep net samples will reveal natural enemies that reside in the tops of the plants, while missing those that live near the base of the plant or in the soil.

- Sampling with traps will only capture mobile natural enemies, but movement can vary with hunger, stage of development, and environmental conditions, and no trap is equally attractive to all natural enemies.

- Sampling by visual searching will detect large and/or relatively immobile natural enemies, while smaller, flightier, or more cryptic natural enemies are likely to be overlooked.

- Sampling field edges and interiors can give you different results. Natural enemies are often more abundant in field edges than interiors, but this difference is especially pronounced early in the season as they colonize the field from overwintering sites. Numbers in the interior will probably increase over time as long as you minimize disruptive practices. If pests are abundant and natural enemies are scarce in the field interior, try sampling field margins to assess natural enemy populations since the last major disturbance (such as planting, tillage, pesticide application, etc.).

Assessing the success of a biological control program

Just as you need to scout repeatedly to monitor pest populations, you also need to continually monitor the level of biological control of those pests. However, detecting and quantifying biological control in the field is seldom as straightforward a task as counting pests, since the activity of natural enemies is seldom as obvious as the presence of a pest. For example, a predator may consume the entire body of its prey, leaving no sign of its activity. Also, a recently parasitized or infected pest may appear healthy though it will inevitably die later. In addition, for a given pest, you may need to monitor for different natural enemies depending on the life stage of the pest (egg, larva, adult), the stage of the crop, and the geographic location. The complicated nature of these and other factors means that few research-based economic thresholds have been developed to predict the level of natural enemy activity needed to reduce pest numbers or prevent economic injury to the crop.

Nevertheless, there are some ways that you can investigate the relative abundance and activity of natural enemies in a particular situation. None of the sampling methods described below will provide a complete picture of natural enemies (see textbox), either in terms of absolute abundance or identifying all species and stages present, but useful information can still be gained by using a given sampling method over multiple years in a given location, especially if done in conjunction with monitoring of pest numbers and good record-keeping. These methods may be most useful in informally evaluating a "biological control experiment," such as an attempt to conserve, enhance, or augment natural enemies, provided that the "experiment" includes areas where the new methods were not used, allowing comparisons of pest and natural enemy activity.

Visual field counts

Visual examination of crop plants in the field is a relatively simple way to assess natural enemy abundance, except for the most mobile and easily disturbed species. It also has the advantage of not requiring any special equipment. Begin by choosing an appropriate unit of the crop to sample. This will vary by size and structure of the crop plant or habitat, but it might be something like 10 feet of row, 10 consecutive plants, 10 branches, 1 square yard, or some similar measure. Alternatively, the sample unit could be a particular length of time, such as 5 minutes. Whatever sample unit you choose, begin by approaching the sample unit slowly, watching for any natural enemies that might fly or run away. Then examine the area as thoroughly as possible without touching it, so that you might observe any easily disturbed insects on the periphery of the sample unit. Finally, complete your sample by moving the plant material as necessary to examine the entire sample unit.

Sweeping

For some crops, sweep net samples can be used to quantify the abundance of both pests and natural enemies. A good way to hone your skills at recognizing natural enemies is to examine the contents of a sweep sample from an unsprayed alfalfa field, which generally harbors a wide variety of pests and natural enemies. Take 10 to 50 sweeps of an alfalfa field with a standard sweep net, then chill the trapped organisms to slow down their movement for easier observation. The easiest way to do this is to shake the net contents to the bottom of the net and place the bottom of the net into a cold ice chest, close the lid, and chill the insects for about 10 minutes. Alternatively, you can attempt to transfer the contents of the net to a plastic bag or other sealable container that can be placed in a cooler or refrigerator for later examination. You can also transfer the contents to a plastic bag or container and freeze it for later examination; this may be more convenient, but you lose the ability to observe the characteristic postures, movements, and behaviors of the live organisms, which can boost your ability to recognize these organisms in the field in the future. If you choose to examine live specimens, dump them into a shallow light colored tray (chilled, if possible) or onto a white sheet once they are sufficiently chilled. You will have several minutes to examine the organisms before they warm up and start to walk or fly away. You'll probably find numerous species of immature and adult generalist predators, parasitic wasps, spiders, and numerous species that are not important as either pests or natural enemies. Sweeping at night may reveal the presence of a different complex of natural enemies, since many are nocturnal.

Pitfall traps

Ground-dwelling natural enemies can be monitored using pitfall traps. For each trap, dig a small hole large enough to sink a plastic cup (16 oz cups are ideal) so that its rim is even with the soil surface. You'll need to place about 2 cm of a killing agent into the bottom of each trap to prevent the trapped organisms from escaping and/or eating each other. A weak dish soap solution (2–3 drops in a quart of water) will suffice if you will be able to check the trap in 24 hours. Organisms left in the soap water longer than that will begin to decompose, so for longer intervals between checking the traps, you need a killing agent with some preservative properties; we recommend a 1:1 solution of nontoxic antifreeze (*propylene* glycol) and water. Do not use regular antifreeze (*ethylene* glycol) as it is attractive and poisonous to vertebrates, including dogs. You can also suspend a board over the trap, an inch or two above the soil surface, to prevent the trap from flooding in case of rain. To examine the trapped organisms, dump the trap's contents into a shallow white pan or tray and sort through them. You'll probably find large numbers of ground beetles, rove beetles, and spiders, along with plenty of other soil organisms.

Yellow sticky traps

Yellow sticky traps, such as those available for use in greenhouses, are attractive to a variety of flying adult insects, including many parasites. They are not equally attractive to all natural enemies; a great many other natural enemies, including immatures and less mobile species will not be captured. The best use of sticky traps is to compare the relative abundance of certain mobile natural enemies between locations. For example, you can use them to gain a sense of how natural enemies are distributed in the edges and centers of crop fields and in noncrop habitats such as old fields, insectary plants, grassy fields, and so on. Sticky traps give the best estimates of natural enemy abundance when left in place for 48 hours.

Assessing parasitism of plant pests

While many sampling methods can detect the presence of adult parasites, the important measure of parasite activity is the percentage of pests that are parasitized. Determining this is not straightforward, since parasitized insects may look healthy for a while as the parasite develops. The best way to determine percent parasitism is to collect host insects from the field, rear them in confinement, and watch for parasite emergence. To do this, you need to know enough about the particular pest to supply its needs for up to 2 weeks, and enough about its natural enemies to collect the life stage(s) of the pest that are likely to be parasitized. Exactly how to do the rearing will depend on the pest and natural enemies being reared, but the following general suggestions should get you started. For more details on rearing and identification, see Gibb and Oseto in the suggested readings list.

If potted host plants are available, place the collected pest insects on a potted plant and use a fine mesh sleeve (e.g., fine mosquito netting) supported by stakes to enclose the insects. Tie the sleeve tightly around the rim of the pot and at the top of the sleeve.

If potted host plants are not available, insert the cut ends of sprigs of the host plant (unsprayed, of course!) in a florist's water pick, prescription vial, or other small container. This will keep foliage in suitable condition, while reducing the likelihood of insects drowning as they might in a larger container such as a vase or jar. Place these cuttings in a clear container containing a few inches of slightly dampened sand, then cover the opening with fine nylon mesh. The damp sand will increase humidity within the container and provide a suitable pupation site for insects that pupate in soil, whereas the mesh will allow adequate ventilation and prevent small specimens from escaping. Including a crumpled paper towel in the container will provide an alternate pupation site for insects. Replace foliage and remove accumulated frass (insect excrement) as needed, being careful neither to introduce new natural enemies with the fresh foliage nor to discard host insects or parasites with the old plant material.

Regardless of the plant material you are using, keep the rearing containers in a lighted location but out of direct sunlight. Record the number of host insects initially placed in each cage and how many hosts develop into adult hosts, yield parasites, or die from disease or other causes. When interpreting the results, keep in mind that for some parasites, several to many individual parasites can emerge from one host. When multiple parasites emerge from a single host, they are usually all the same species, but it is possible for two species of parasite to emerge from a single host. Thus, if you want to determine the percentage of parasitism, you really need to keep track of what happens to the hosts, rather than simply count the number of adult parasites you find in the rearing container.

Assessing filth fly parasitism

For filth fly control, the best measure of effectiveness is a reduction in the number of adult flies. However, you may want to assess the contribution of fly parasites to overall fly control, especially if you are making augmentative releases of filth fly parasites. Begin by adding a shovelful of manure containing filth fly pupae to a large pail of water. As you stir and agitate the manure and water, the pupae will float to the surface. Remove these pupae for examination. Pupae that are old enough to have been attacked by fly parasites will have turned from reddish brown to a darker brown. Pupal cases in which the end cap is neatly broken off have already released adult filth flies. Pupae that have neat circular holes somewhere other than the end cap have already yielded adult parasites. The remaining fly pupae with no holes can be reared in a jar with a damp paper towel for several days to allow the insects to complete their development and emerge. You can determine the percent parasitism by examining 100 pupae and counting the number that yielded adult parasites. Keep in mind that percent parasitism can underestimate the parasites' actual impact, since host-feeding by the adult parasites could have killed at least some of the pupae that did not produce adult insects.

The best advice: Start small and learn as you grow

When adopting any new practice, a small mistake can sometimes have significant consequences. The best way to gain experience and confidence using biological control methods is to start on a small scale. Try these new methods on one plot or field while continuing to use your usual practices on the rest of the crop. Your experimental area should be large enough so that the results will reflect your normal crop management practices with the additional effects of natural enemies. It should also be large enough that predators and parasites will be unaffected by sprays or other disruptive practices from outside the experimental area. However, the experimental area should be small enough that mistakes will not be excessively costly. Also, it should be conveniently located to facilitate frequent monitoring of pest and natural enemy activity throughout the trial period.

Spend time identifying natural enemies, watching their activities, and assessing their impact on the pest population. It is natural to be reluctant to entrust the safety of your crop to an assemblage of insects and microorganisms. If you spend some time learning to recognize natural enemies and assessing their activity in your own field, you will become more confident in the significant contributions they make to your pest management program.

After you assess the benefits in your test areas, you will be ready to apply the practices on a larger scale. Expand only as rapidly as your levels of comfort and anxiety allow. Always remember that frequent and routine monitoring is essential to the success of any pest management program. Finally, if you have questions or concerns about your program, immediately contact your supplier, crop consultant, or Cooperative Extension Service.

Why there are so few specific guidelines for biological control

Ideally, biological control can help growers produce a profitable crop in a sustainable and environmentally sound manner. If a grower knows that enough natural enemies are present to reduce pest numbers to tolerable levels, the grower can avoid an insecticide application and profit from the decision, immediately and in the long run. If natural enemies will not provide adequate control, knowing which ones were present and what they were doing will not compensate for low yields or low quality and an unprofitable crop. Unfortunately, in most situations we simply do not know how many natural enemies are needed to provide sufficient pest control.

One exception is with European red mites and their predators in apples. For many years, research-based thresholds have been available to determine if natural enemies will adequately control European red mites or if a miticide application is necessary. These thresholds are based on densities of the pest and ratios of predators (the lady beetle *Stethorus punctum* and the phytoseiid mite *Amblyseius fallacis*). We need this type of information for many more pests and their natural enemies.

Although guidelines for biological control of alfalfa weevil are not as detailed as those for European red mite, they are more specific than those that have been developed for most other pests. Biological control of alfalfa weevil highlights the challenges behind developing specific threshold-based biological control recommendations for more pests. Several parasites of alfalfa weevil have been established through classical biological control. These parasites provide adequate control in some situations but not in others. In general, biological control of alfalfa weevil is more often successful in northern parts of the Midwest than in the South. Some information is available on how specific ratios of parasites to weevils affect weevil abundance, but these have not been formulated into grower-friendly guidelines for forecasting levels of biological control of alfalfa weevil in a given field. In addition, it is known that biological control of alfalfa weevil tends to be relatively stable from year to year in the absence of insecticide use, but we do not know the exact effects on weevil biological control of an insecticide application for another pest, such as potato leafhopper. On the other hand, an Iowa Extension publication, *Biological Control of the Alfalfa Weevil in Iowa* (www.extension.iastate.edu/ Publications/PM1484.pdf), advises farmers to conserve alfalfa weevil parasites by avoiding insecticide applications from May 1 to 10, when the adult wasps are most active. In addition, the publication explains how early harvest of first cutting alfalfa can promote infection of alfalfa weevil by *Zoophthora phytonomi*, an important fungal pathogen.

For most other crops and pests, we simply do not have the information needed to relate pest and natural enemy abundance to expected level of biological control. In part, this lack of knowledge is because the information is so difficult to generate. Also, seemingly subtle differences in crop management (such as planting date or row spacing), pest management (such as weed control or choice of insecticide options), or the physical environment (such as weather or soil type) may have profound effects of important natural enemies, either singly or in combination. Although we proclaim that agricultural crops are simple ecosystems (and they are in comparison with natural ecosystems), in actuality they are sufficiently complex to often hamper accurate predictions of natural enemy impacts. Nevertheless, wider adoption of biological control will depend, in part, on its predictability, and we hope that biological control researchers will continue to pursue the development of additional threshold-based biological control programs like those described above.

Additional reading

Numerous resources on specific topics are mentioned throughout the text. More general resources that you may find helpful include the following:

Print resources

Altieri, M.A., C.I. Nicholls, and M.A. Fritz. 2005. *Manage Insects on Your Farm: A Guide to Ecological Strategies.* Sustainable Agriculture Network Handbook No. 7. 119 pp. Principles and practical examples of conservation biological control.

Barbosa, P. (editor). 1998. *Conservation Biological Control.* Academic Press. 396 pp. An academic review of research on conservation of natural enemies.

Cranshaw, W. 2004. *Garden Insects of North America: The Ultimate Guide to Backyard Bugs.* Princeton University Press. 672 pp. An amply illustrated guide to identification of garden pests and natural enemies, together with more background on insect biology and management.

Hajek, A. 2004. *Natural Enemies: An Introduction to Biological Control.* Cambridge University Press. 394 pp. An undergraduate-level textbook covering biological control of arthropods, vertebrates, plant pathogens, and weeds.

Hoffman, M.P. and A.C. Frodsham. 1993. *Natural Enemies of Vegetable Insect Pests.* Cornell Cooperative Extension. 63 pp. Covers insect biology and ecology, principles of IPM and biological control, and important natural enemies of vegetable insect pests.

Flint, M.L. and S.H. Driestadt. 1998. *Natural Enemies Handbook: The Illustrated Guide to Biological Pest Control.* University of California Press. 154 pp. An illustrated and detailed guide to biological control and natural enemies with a strong emphasis on California.

Gibb, T.J. and C.Y. Oseto. 2006. *Arthropod Collection and Identification: Laboratory and Field Techniques.* Elsevier. 311 pp. Strategies for collecting, rearing, and identifying insects.

Johnson, W.T., and H.H. Lyon. 1991. *Insects That Feed on Trees and Shrubs.* Cornell University Press, 560 pp. A great resource for identifying pests.

Mahr, S.E.R., R.A. Cloyd, D.L. Mahr, and C.S. Sadof. 2001. *Biological Control of Insects and Other Pests of Greenhouse Crops.* North Central Regional publication 581. 100 pp. Detailed information on pests, natural enemies, and implementation of biological control in greenhouses. Available free online or for purchase at learningstore.uwex.edu

Mahr, S.E.R., D.L. Mahr, and J.A. Wyman. 1993. *Biological Control of Insect Pests of Cabbage and Other Crucifers.* North Central Regional publication 471. 54 pp. Detailed information on pests, natural enemies, and implementaion of biological control in brassicas. Available free online at learningstore.uwex.edu

McCullough, D.G., S. Katovich, D. Neumann, D. Mahr, C. Sadof, and M. Raupp. 1999. *Biological Control of Insect Pests in Forested Ecosystems: A Manual for Foresters, Christmas Tree Growers and Landscapers.* Michigan State University publication E-2679. 123 pp. A research-based guide to the important pests in forests and tree plantations, their natural enemies, and approaches to biological control in such systems. Available free online at web2.msue.msu.edu/bulletins/Bulletin/PDF/E2679.pdf

Olkowski, W., S. Daar, and H. Olkowski. 1991. *Common-Sense Pest Control: Least-Toxic Solutions for Your Home, Garden, Pets and Community.* Taunton Press. 715 pp. Includes some biological control approaches among other least-toxic solutions for a wide variety of pest problems.

Westcott, C. 1973. *The Gardener's Bug Book, 4th ed.* Doubleday. 689 pp. Very useful for identification and life histories of garden insects, but the control recommendations are seriously out-of-date.

Internet resources

2002 Directory of Least-Toxic Pest Control Products. 2001. IPM Practitioner Vol. XXIII No. 11/12. Lists suppliers and commercially available products, services, and natural enemies for use in integrated pest management and biological control. The 2002 edition is available free online; a more current directory may also be available for purchase.
www.birc.org/products.pdf

ATTRA—National Sustainable Agriculture Information Service. Free, in-depth publications on innovative production and marketing practices, organic certification, and highlights of local, regional, USDA and other federal sustainable agriculture initiatives. Many of these publications include specific applications of biological control.
www.attra.org

Biological Control: A Guide to Natural Enemies in North America. Provides photographs, descriptions, and life cycle information for over 100 natural enemies of insect, disease, and weed pests in North America and a brief tutorial on the concept and practice of biological control and integrated pest management.
www.nyaes.cornell.edu/ent/biocontrol

Midwest Biological Control News. The archives of this monthly newsletter (no longer published) provide a wealth of information on biological control in farm, garden, and home settings. The archives are not searchable, but a comprehensive index can help locate the desired information.
www.entomology.wisc.edu/mbcn/mbcn.html

Appendix 1

Common natural enemies of major insect pests of important crops in the North Central United States

Alfalfa

Alfalfa weevil

Parasites: *Anaphes luna* and *Tetrastichus incertus* (Chalcidoidea), *Bathyplectes anurus* and *B. curculionis* (Ichneumonidae), and *Microctonus aethiopoides* and *M. colesi* (Braconidae)

Predators: predatory bugs

Pathogens: *Zoophthora phytonomi*, *Beauveria bassiana**

Aphids

Aphid generalists

Aphid generalists

Parasites: various parasitic wasps,* including some in the genera *Aphidius,** *Aphelinus,** *Lysiphlebus,** *Praon,** and others (for augmentation, you must identify the aphid species)

Predators: smaller predatory Hemiptera (especially *Orius* spp.*), lacewings,* lady beetles,* *Aphidoletes* gall midges,* ground beetles, hover flies; aphids that drop to the soil often are attacked by spiders, ground beetles, and rove beetles

Pathogens: *Beauveria bassiana,** *Metarhizium anisopliae,** *Paecilomyces fumosoroseus**

Caterpillar generalists

Parasites: tachinid flies, *Trichogramma* spp.* and other wasps* (for augmentation, you must identify the caterpillar species)

Predators: predatory Hemiptera,* lacewings,* lady beetles,* ground beetles, yellowjackets, hornets

Pathogens: *Bacillus thuringiensis* ssp. *kurstaki* (Btk),* *Bacillus thuringiensis* ssp. *aizawai* (Bta),* *Beauveria bassiana**

Potato leafhopper

Parasites: *Anagrus nigriventris* (Chalcidoidea)

Predators: aphid generalists (especially damsel bugs and *Orius* spp.*), orb web spiders

Pathogens: *Zoophthora* (= *Erynia*) *radicans*

Tarnished plant bug

Parasites: *Peristenus digoneutis* (Braconidae) (expanding slowly westward after introduction in New Jersey in 1984), *Leiophron* spp. (Braconidae), *Anaphes iole** (Chalcidoidea), other parasitic wasps, tachinids

Predators: bigeyed bugs,* damsel bugs, predatory plant bugs

Pathogens: *Beauveria bassiana**

Berries

Leafrollers and other caterpillars

Caterpillar generalists

Parasites: *Macrocentrus ancylivorus** (Braconidae)

Pathogens: insect-pathogenic nematodes* may be effective against strawberry leafrollers

Spider mites

Predators: minute pirate bugs,* *Phytoseiulus persimilis,** *Neoseiulus californicus** (Phytoseiidae)

Spittlebugs

No good natural enemies

Tarnished plant bug

See listing under alfalfa

Weevils and raspberry fruitworm

Predators: ground beetles, rove beetles

Pathogens: insect-pathogenic nematodes* may be effective for some strawberry pests, *Beauveria bassiana**

Brassicas (cabbage and relatives)

See *Biological Control of Insect Pests of Cabbage and Other Crucifers* (NCR471).

Cabbage aphids

Aphid generalists

Parasites: *Diaeretiella rapae** (Braconidae)

Imported cabbageworm

Caterpillar generalists

Parasites: *Pteromalus puparum** (Chalcidoidea), *Cotesia glomerata* and *C. rubecula* (Braconidae)

Pathogens: *Artogeia rapae* granulosis virus (ARGV)

Cabbage looper

Caterpillar generalists

Parasites: *Cotesia marginiventris* (Braconidae);* *Copidosoma floridanum* (Chalcidoidea), *Phryxe vulgaris* and *Voria ruralis* (Tachinidae)

Pathogens: cabbage looper NPV

Diamondback moth

Caterpillar generalists

Parasites: *Cotesia plutellae** and *Microplitis plutellae** (Braconidae), and *Diadegma insulare** and *Diadromus* spp. (Ichneumonidae)

Flea beetles

Parasites: *Microctonus vittatae* (Braconidae)

Predators: various generalist predators have limited effect

Pathogens: *Beauveria bassiana,** insect-pathogenic nematodes* attack larvae in soil with limited effect

Cabbage maggot

Parasites: *Trybliographa rapae* (Cynipoidea), *Aleochara bilineata* (Staphylinidae)

Predators: ground beetles, rove beetles, larval soldier beetles

Pathogens: insect-pathogenic nematodes*

---

* commercially available natural enemies for augmentation

Corn

Aphids

Aphid generalists

Corn earworm

Caterpillar generalists, especially minute pirate bugs*

Parasites: *Cotesia marginiventris* and *Microplitis croceipes* (Braconidae)

Predators: *Coleomegilla maculata* (Coccinellidae)

Corn rootworms

Parasites: *Syrrhizus diabroticae* (Braconidae)

Predators: acarid mites, ants, ground beetles, rove beetles, larval soldier beetles

Pathogens: insect-pathogenic nematodes*

European corn borer

Caterpillar generalists, especially minute pirate bugs* and lady beetles

Parasites: *Lydella thompsoni* and other tachinids, *Macrocentrus grandii* (Braconidae), *Eriborus terebrans* (Ichneumonidae),

Pathogens: *Nosema pyraustae* (a protist)

Seed corn maggot

Predators: ground beetles, rove beetles, larval soldier beetles

Pathogens: some naturally occurring fungi, insect-pathogenic nematodes*

Cucurbits (squash, melons, cucumbers)

Squash vine borer

Pathogens: *Bacillus thuringiensis* ssp. *kurstaki* (Btk)* and *Bacillus thuringiensis* ssp. *aizawai* (Bta)* if present on stems when newly hatched larvae enter stems; insect-pathogenic nematodes* or Bta* or Btk* may be injected into stems containing borers

Squash bug

Parasites: *Gryon pennsylvanicum* and *Eumicrosoma* spp. (Scelionidae), *Trichopoda pennipes* (Tachinidae)

Predators: spiders, ground beetles, and other generalists

Cucumber beetles

Parasites: *Celatoria diabroticae* (Tachinidae), braconid wasps

Predators: generalist predators in soil may attack larvae and pupae

Pathogens: insect-pathogenic nematodes* may be used against larvae and pupae in the soil

Filth flies and related livestock pests

Parasites: *Muscidifurax* spp.,* *Nasonia vitripennis*,* and *Spalangia* spp.* (Chalcidoidea)

Predators: hister beetles*

Greenhouse pests

See *Biological Control of Insects and Other Pests of Greenhouse Crops* (NCR581) for more details

Aphids

Aphid generalists

Fungus gnats and shore flies

Predators: *Hypoaspis miles** (Laelapidae)

Pathogens: insect-pathogenic nematodes,* *Bacillus thuringiensis* ssp. *israelensis* (Bti)*

Agromyzid leafminers

Parasites: *Dacnusa sibirica** (Braconidae), *Diglyphus isaea** (Chalcidoidea)

Pathogens: *Steinernema* (=*Neoaplectana*) *carpocapsae** (insect-pathogenic nematode)

Scales and mealybugs

Many pest species exist and many natural enemies are available for augmentation, see NCR581

Spider mites (must identify species first)

Predators: *Hypoaspis aculeifer*,* *Galendromus* (=*Metaseiulus*, =*Typhlodromus*) *occidentalis*,* *Mesoseiulus* (=*Phytoseiulus*) *longipes*,* *Neoseiulus* (=*Amblyseius*) *californicus*,* *Phytoseiulus macropilis** and *P. persimilis** (all Phytoseiidae); *Feltiella acarisuga** (Cecidomyiidae); *Scolothrips sexmaculatus** (Thripidae); *Stethorus* spp. (Coccinellidae)

Whiteflies (must identify species first)

Parasites: *Encarsia formosa*,* *E. luteola*,* *Eretmocerus californicus** (Chalcidoidea)

Predators: green lacewings,* bigeyed bugs,* minute pirate bugs,* *Delphastus pusillus** (Coccinellidae)

Pathogens: *Beauveria bassiana**

Thrips (must identify species first)

Parasites: *Thripobius semiluteus** (Chalcidoidea)

Predators: *Orius* spp.* (Anthocoridae), *Hypoaspis miles** (Laelapidae), *Neoseiulus* (=*Amblyseius*) *cucumeris** and *N. degenerans** (Phytoseiidae),

Pathogens: *Beauveria bassiana*,* *Metarhizium anisopliae**

Landscape pests

The large number of plants and pests preclude listing them here. Begin by identifying your pest; a great resource for this is *Insects That Feed on Trees and Shrubs*. Knowing the identity of your pest will help you begin to track down information on natural enemies.

Small grains

Aphids (including greenbug)

Aphid generalists, especially *Hippodamia convergens*,* *Coccinella septempunctata* and *Coleomegilla maculata** (Coccinellidae), and syrphids

Parasites: various braconid and aphelinid wasps, especially *Diaeretiella rapae** and *Lysiphlebus testaceipes** (Braconidae), *Aphelinus albipodus* and other *Aphelinus* spp.* (Aphelinidae)

Cereal leaf beetle

Parasites: *Anaphes flavipes** and *Tetrastichus julis* (Chalcidoidea)

Predators: lady beetles*

Grasshoppers

Parasites: blister beetle larvae, various tachinid flies

Predators: generalist predators

Pathogens: *Beauveria bassiana*,* *Nosema locustae**

Solanums (tomato, potato, eggplant, and relatives)

Aphids

Aphid generalists

Colorado potato beetle

Parasites: *Myiopharus* (=*Doryphorophaga*) spp. (Tachinidae), *Edovum puttleri*

Predators: ground beetles (esp. *Lebia grandis*), *Coleomegilla maculata** (Coccinellidae), predatory stink bugs*

Pathogens: *Bacillus thuringiensis* ssp. *tenebrionis* (Btt),* *Beauveria bassiana**

Flea beetles

Predators: various generalist predators have limited effect

Pathogens: insect-pathogenic nematodes* attack larvae in soil with limited effect

Soybean

Soybean aphid

Aphid generalists, especially lady beetles, minute pirate bugs, lacewings, and syrphids

Parasites: *Lysiphlebus testaceipes*,* *Aphidius colemani** and *Binodoxys communis* (Braconidae), and many other wasp species

Pathogens: *Pandora neoaphidis* (a fungus)

Bean leaf beetle

Parasites: *Trombidium* mites, *Celatoria diabroticae* and *Medina* sp. (Tachindae)

Predators: ground beetles, predatory Hemiptera

Pathogens: *Beauveria bassiana*,* *Metarhizium* sp.

Stored products

Moth larvae

Parasites: *Bracon hebetor** and *Cotesia marginiventris** (Braconidae), *Trichogramma* spp.*

Predators: *Lyctocoris campestris* and *Xylocoris flavipes** (Anthocoridae)

Pathogens: *Bacillus thuringiensis* ssp. *aizawai* (Bta)* and *Bacillus thuringiensis* ssp. *kurstaki* (Btk)*

Grain beetles

Parasites: *Anisopteromalus calandrae** (Chalcidoidea)

Predators: *Lyctocoris campestris* and *Xylocoris flavipes** (Anthocoridae)

Tree fruit

Plum curculio

Parasites: several wasps attack eggs and larvae with limited effectiveness

Predators: generalist predators in soil attack larvae and pupae with limited effectiveness

Pathogens: *Beauveria bassiana*,* insect-pathogenic nematodes*

Codling moth

Caterpillar generalists

Parasites: *Ascogaster quadridentata* (Braconidae)

Pathogens: codling moth GV*

Aphids

Aphid generalists

Leafrollers

Caterpillar generalists

Parasites: *Actia interrupta* (Tachinidae), *Macrocentrus ancylivorus* (Braconidae)

Leafminers

Parasites: *Pholetesor* spp. (Braconidae), *Sympiesis* spp. (Chalcidoidea), and numerous other wasps

Predators: lacewing larvae

* commercially available natural enemies for augmentation

Appendix 2
A simplified key for identifying common natural enemies

The following key, in combination with the photos and diagrams in this book, can be used to begin to identify common natural enemies. It will work best if you have good reason to believe that the insect or spider you want to identify is in fact a natural enemy. Given the diversity of natural enemies and other arthropods, this simple key may not yield correct identifications for every arthropod you attempt to identify, especially if it is not a natural enemy. Indicated sizes are for the more commonly encountered parasitic or predatory species.

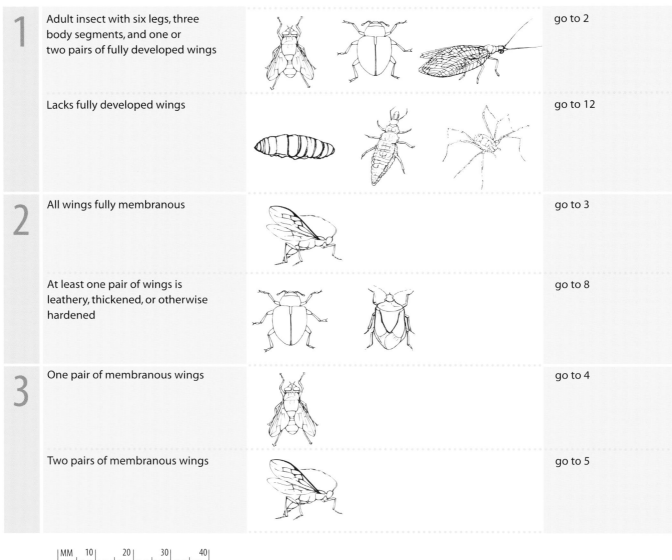

1	Adult insect with six legs, three body segments, and one or two pairs of fully developed wings		go to 2
	Lacks fully developed wings		go to 12
2	All wings fully membranous		go to 3
	At least one pair of wings is leathery, thickened, or otherwise hardened		go to 8
3	One pair of membranous wings		go to 4
	Two pairs of membranous wings		go to 5

| MM 10 20 30 40 |
| INCH   1 |

4 **Order Diptera
(flies, gnats, midges, others)***

Body with markings like a bee
or wasp, 5–15 mm

**Family Syrphidae:
Hover flies**
(see page 39)

Body black to gray, with or without stripes, like a house fly but
with stout bristles on abdomen,
5–12 mm

**Family Tachinidae:
Tachinid flies**
(see page 40)
—if striped, could
be Sarcophagidae
(flesh flies)

Body metallic blue, green or
copper-colored, with long slender
legs, 3–7 mm

**Family
Dolichopodidae:
Long-legged flies**
(see page 41)

Body long, tapered, and covered
with fine hairs, 10–30 mm

**Family Asilidae:
Robber flies**
(see page 41)

5 Front wings only slightly larger
than hind wings, wings with
extensive netted venation, no
narrow waist between thorax
and abdomen

**Order Neuroptera:
Lacewings, others**
go to 6

Front wings noticeably larger than
hind wings, wing venation moderate
to nearly absent, thorax joined to
abdomen by a narrow waist

**Order Hymenoptera:
Wasps, bees, ants**
go to 7

6 **Order Neuroptera
(lacewings and others)***

Body bright green, 10–15 mm

**Family Chrysopidae:
Green lacewings**
(see page 36)

Body brown, 6–12 mm

**Family Hemerobiidae:
Brown lacewings**
(see page 37)

*Use the descriptions, sizes, and photographs above to identify further.

7 Order Hymenoptera (wasps, bees, ants)*

10–30 mm		**Family Vespidae: Stinging wasps** (see page 48)
10–30 mm		**Family Sphecidae: Stinging wasps** (see page 48)
2–12 mm		**Family Braconidae: Braconid wasps** (see page 43)
2–20 mm		**Family Ichneumonidae: Ichneumonid wasps** (see page 44)
0.25–4 mm		**Superfamily Chalcidoidea: Chalcidoid wasps** (see page 45)
One or two camel-like humps on the waist between the thorax and abdomen, 3–15 mm		**Family Formicidae: Ants**, winged form (see page 48)

8

Front wings uniformly hardened, meet along midline of abdomen and entirely cover the membranous hind wings		**Order Coleoptera: Beetles** go to 9
Front wings at least partially leathery or thickened but not as described above		go to 10

*Use the descriptions, sizes, and photographs above to identify further.

9 ## Order Coleoptera (beetles)*

Body and front wings black, brown, or metallic blue or green; wings have uniform grooves or rows of dimples; abdomen usually wider than thorax and head; usually found on soil or under debris; 3–40 mm

Family Carabidae Ground beetles
(see page 30)

Body slender; body and wings black or dark brown; wings very short, leaving much of abdomen exposed; usually found on soil or under debris; 2–10 mm

Family Staphylinidae: Rove beetles
(see page 34)

Body round to oval, convex; front wings smooth and shiny, usually red, orange, pink, or black and often with spots of a different color; usually found on foliage; 2–8 mm

Family Coccinellidae: Lady beetles
(see page 31)

10 Front wings uniformly leathery; first thoracic segment much longer than head; front legs modified for grasping prey; chewing mouthparts; 50–80 mm

Order Mantodea: Praying mantids
(see page 26)

Front wings thickened and leathery near base and membranous with venation visible near tip; stylet-like mouthparts held along underside of thorax; 1–15 mm

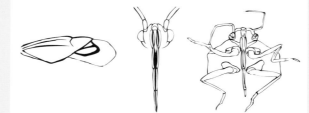

Order Hemiptera: True bugs
go to 11

*Use the descriptions, sizes and photographs above to identify further.

MM 10 20 30 40
INCH 1

11 Order Hemiptera (true bugs)*

Body 1.5–2x longer than wide, shield-shaped; head as wide as long but much narrower than thorax; 10–15 mm

Family Pentatomidae: Stink bugs
(see page 29)

Body elongate, about 5x longer than wide; gray to tan; front legs with thicker femur than other legs; 6–10 mm

Family Nabidae: Damsel bugs
(see page 29)

Body 3–4x longer than wide; middle part of abdomen often sticks out laterally beyond wings; brown, black, or irregularly patterned; front legs with thicker femur than other legs; some may be ambush hunters on flowers;10–20 mm

Family Reduviidae: Assassin bugs and ambush bugs
(see page 28)

Body about 3x longer than wide; tan, brownish, or black and white; large eyes protrude laterally beyond thorax, 3–4 mm

Family Lygaeidae: Seed bugs and bigeyed bugs
(see page 28)

Body flattened, teardrop-shaped, and 2–3x longer than wide; front wings with distinctive black and white markings; front wings more or less in a single plane, 1–2 mm

Family Anthocoridae: Minute pirate bugs
(see page 27)

Body flattened and 2–3x longer than wide; brown with tan markings; membranous part of front wings not in a plane with the leathery part but folded downward at a distinct angle; pronounced notch between leathery and membranous parts of front wings; 2–4 mm

Family Miridae: Plant bugs
(see page 27)

MM 10 20 30 40
INCH 1

*Use the descriptions, sizes, and photographs below to tentatively identify further; unless specified, usually found on foliage.

12	No apparent legs, maggotlike body		**Order Diptera: Fly larvae** go to 13
	Three pairs of jointed legs, body insect-like with three regions		go to 14
	Four pairs of jointed legs		go to 16

13 Order Diptera (fly larvae)*

| | Body more or less cylindrical and orange-to-red in color, <4 mm | | **Family Cecidomyiidae: Gall midge larvae** (see page 38) |
| | Body tapered, narrowest near head and widest near tip of abdomen; ivory, yellow, green, or brown, often with colorful markings; 2–15 mm | | **Family Syrphidae: Hover fly larvae** (see page 39) |

| 14 | Found in soil or litter | | **Family Carabidae: Ground beetle larvae, Family Staphylinidae: Rove beetle larvae,** or other immature insect (see pages 30 and 34) |
| | Found on foliage | | go to 15 |

*Use the descriptions, sizes, and photographs above to tentatively identify further; unless specified, all are usually found on foliage.

15   Stylet-like mouthparts held along underside of thorax while at rest, or extended forward for feeding; small flap-like wing buds

Because this order has simple metamorphosis, nymphs can be tentatively identified by resemblance to adults

Order Hemiptera: True bug nymphs (see pages 26–29)

Body alligator-like but not particularly "warty;" sickle-like mandibles extend forward from head; gray to brown, often with white or ivory markings; 2–15 mm

Chrysopidae: head similar in width to first segment of thorax
Hemerobiidae: head narrower than first segment of thorax

Order Neuroptera: Lacewing larvae (see pages 36–37)

Body alligator-like, usually "warty" or spiny; dark gray to black, often with white, yellow, or orange spots; 2–10 mm

Family Coccinellidae: Lady beetle larvae (see pages 31–33)

16   Body has two regions: a cephalothorax, to which the legs are attached, connected by a thin waist to the abdomen

Order Araneida: Spiders (see pages 49–50)

Body has just a single region

go to 17

17   Body 4–7mm long; stilt-like legs much longer than body

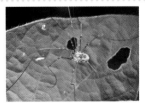

Order Opiliones: Harvestmen (see page 51)

Body minute, nearly invisible without magnification; legs no longer than body

Order Acari: Mites (see page 51)

Glossary

Abdomen. The third body region of an insect, behind the head and thorax. Contains many of the digestive and reproductive organs.

Abiotic factors. Environmental conditions not resulting from living organisms. Important in natural control. Includes weather and soil conditions. See biotic factors; natural control.

Action threshold. Pest population level at which control measures should be initiated to prevent undesirable damage to the crop. See economic injury level; economic threshold.

Arthropod. A member of the animal phylum that includes insects, spiders, ticks, mites, centipedes, millipedes, and crustaceans. The name means jointed legs, but these animals also have an exoskeleton and must molt in order to grow larger.

Augmentative biological control. One of three general approaches to biological control. It is the periodic release of captured or artificially produced natural enemies to supplement those that occur naturally. Also called augmentation of natural enemies.

Biological control. The management of pest populations by the purposeful manipulation of beneficial natural enemies.

Biotic factors. Environmental factors related to living organisms. Important in natural control. Includes predation, parasitism, disease, and competition for resources. See abiotic factors; natural control.

Bt. Abbreviation for the bacterial insect pathogen *Bacillus thuringiensis*.

Classical biological control. One of three general approaches to biological control. A process in which scientists seek effective beneficial natural enemies in the native home of the pest, and introduce and permanently establish these natural enemies in the pest's new habitat, thereby approximating permanent natural control. Also called importation of natural enemies.

Complete metamorphosis. Type of insect development in which the insect passes through a series of larval stages and a pupal stage before becoming an adult. See simple metamorphosis.

Conservation biological control. One of three general approaches to biological control. It is the provision of necessary requirements for natural enemies and the avoidance of practices that kill them or interfere with their beneficial activities. Also called conservation of natural enemies.

Cuticle. The outer body covering of insects, mites, and other arthropods.

Developmental threshold. The lowest temperature at which insect growth and development occurs, usually in the range of 40–50°F.

Direct pest. A pest species that directly feeds on and damages the useful part of the plant. See indirect pest.

Economic injury level (EIL). Pest population level that will result in economic damage if left untreated.

Economic threshold (ET). Pest population level at which control measures should be initiated to prevent the population from exceeding the economic injury level.

Epizootic. A disease outbreak among insects or animals in which a high percentage of the population rapidly becomes infected, often resulting in a high death rate.

Exoskeleton. The external skeleton of an insect, composed of hard cuticle (the "skin").

Generalist natural enemy. A natural enemy that attacks many different types of prey. They often do not have the capability of finding low numbers of a specific pest species. See specialist natural enemy.

Genus (plural is genera). A group of similar and closely related species. A family contains multiple genera, and a genus contains multiple species.

Gregarious parasite. A species in which numerous individual parasites develop within a single host.

Head. The body region in front of the thorax. Contains the brain and bears mouthparts, eyes, antennae.

Hibernation. A prolonged dormant period usually passed in a sheltered location. Many insects use hibernation to wait out periods of adverse environmental conditions. More technical terms for hibernation include diapause (winter dormancy) and aestivation (summer dormancy).

Honeydew. The sweet liquid excretion of sap-feeding insects such as aphids, leafhoppers, and whiteflies. An important alternate or supplemental food for many predatory and parasitic insects.

Host. An organism in or on which a parasite or pathogen lives.

Host feeding. Feeding by the adult female parasite on the same species of insect that serves as a host for her parasitic larvae. Characteristic of certain parasitic flies and wasps.

Host-plant resistance. The defense mechanisms of a plant against a parasite or pathogen. For example, glandular haired varieties of alfalfa are resistant to potato leafhopper. Sometimes called host resistance. See also resistance.

Importation of natural enemies. See classical biological control.

Incomplete metamorphosis. See simple metamorphosis.

Indirect pest. A pest species that feeds on a part of the plant that is not used, but injures the plant sufficiently that yield or quality is reduced. See direct pest.

Inoculation. A preventive method of augmentative biological control in which relatively small numbers of natural enemies are released periodically for sustained management of the pest population. Also called inoculative release. See also augmentative biological control; compare to inundation.

Insectary. A facility for rearing insects. Often refers to companies that mass-produce beneficial predators and parasites for release in augmentative biological control.

Insectary planting. A diverse plot of vegetation planted and managed explicitly for providing natural enemies with resources such as pollen, nectar, shelter, and alternate hosts or prey that might not be available in the adjacent cropping area. Also called habitat planting, conservation buffer, and beneficial insect habitat.

Insect-parasitic nematode. Any of a group of small to microscopic worms that parasitize insects.

Instar. The stage of an immature insect between successive molts. Many insects go through 3–7 instars during their development from egg to adult.

Integrated pest management (IPM). Combining the best of all available methods to control pest populations in the most effective, economic, and environmentally sound manner possible.

Inundation. A curative method of natural enemy augmentation that uses large-scale releases of natural enemies for the immediate reduction of pest populations that are at or near damaging levels. Also called inundative release. See also augmentative biological control; compare to inoculation.

Key pest. See primary pest.

Larva (larvae). The immature form of an insect that undergoes complete metamorphosis; the stage between the egg and pupa. Some common names for larvae include grubs, maggots, and caterpillars. Compare to nymph.

Life cycle. The sequence of life stages from egg to nymph to adult to egg (for insects with incomplete metamorphosis) or from egg to larva to pupa to adult to egg (for those with complete metamorphosis). Sometimes used to mean the *time* required for an insect to complete its life cycle.

Metamorphosis. The series of changes an insect passes through in its growth from egg to adult.

Microbial control. The use of a microbial pesticide to control a pest.

Microbial insecticide. A commercial preparation of living microorganisms (such as bacteria, viruses, or fungi) that cause disease in specific groups of insects. These preparations can be mixed with water and applied with conventional pesticide-application equipment, such as sprayers.

Microbially derived insecticide. A commercial preparation containing toxic compounds derived from microorganisms, but that does not contain viable microorganisms as an active ingredient.

Molt. The process of shedding the skin between developmental stages (instars).

Monoculture. An extensive continuous planting of a single crop.

Natural control. The combined actions of natural environmental events (such as weather, predation, and disease) that reduce pest populations without human involvement.

Natural enemy. A beneficial organism that kills or interferes with pests. See insect-parasitic nematode; parasite; pathogen; predator.

Non-native. An organism found outside its natural or historical range of distribution. In biological control literature, non-native, introduced, alien, exotic, and non-indigenous are used synonymously.

Nontarget organism. An organism that may accidentally be adversely affected by a pest control practice aimed at one or more specific pest species.

Nymph. The immature form of an insect that undergoes simple metamorphosis; the stage between hatching and the winged adult stage. Compare with larva.

Occasional pest. See secondary pest.

Occlusion body. A tough survival package that surrounds the infectious particles (virions) of some insect-pathogenic viruses and can protect them from ultraviolet light and other adverse environmental conditions for several years.

Oviposition. The act of laying or depositing eggs.

Ovipositor. The egg-laying structure of a female insect.

Parasite. An organism that derives its food from the body of another organism (the host). A parasitic insect spends its immature stages in or on the body of a host, which dies just before the parasite pupates.

Parasitoid. A term sometimes used in place of "parasite" for insects that parasitize other insects.

Parasporal body. A crystal of protein toxin produced during the spore-formation process in some insect-pathogenic species of *Bacillus* bacteria, such as *Bacillus thuringiensis*. The toxin in the parasporal body is mostly what makes these bacteria capable of killing their hosts.

Pathogen. An organism, such as a bacteria, fungus, or virus, that is capable of causing disease.

Pest resurgence. A pest outbreak that results from the elimination of the pest's natural enemies, such as after an insecticide application. Even though the pesticide may initially control the target pest, the pest can rebound rapidly because its natural enemies have been eliminated. Compare with secondary pest outbreak.

Pheromone. A chemical substance secreted by an organism for the purpose of communicating with other individuals of the same species.

Polyculture. A mixture of relatively small plantings of different crops in close proximity.

Potential pest. A species that has the potential to cause damage but is usually kept below damaging levels by natural controls.

Predator. An organism that obtains its food by killing two or more prey during its lifetime.

Prey. An organism that a predator kills for food.

Primary pest. A pest species that is present and always damaging unless control measures are taken. Also called key pest.

Proleg. A fleshy, unsegmented abdominal leg on caterpillars and sawfly larvae.

Pupa (pupae). A nonfeeding, inactive stage of complete metamorphosis during which an insect changes from a larva to an adult.

Puparium (puparia). In flies, a protective case formed by the hardening of the larval skin and in which the pupa develops.

Resistance. A genetically determined ability to survive a damaging or lethal factor such as a pesticide. Through natural selection, susceptible populations acquire resistance over many generations of exposure to the damaging or lethal factor. See also host-plant resistance.

Secondary pest. A pest species that is usually present in insufficient numbers to cause economic injury. Also called occasional pest.

Secondary pest outbreak. A rapid increase in the numbers of one pest after treatment for another type of pest. The increase may result from the elimination of the natural enemies of the secondary pest, such as after an insecticide application. Compare with pest resurgence.

Simple metamorphosis. Type of insect development in which the insect does not pass through a pupal stage before becoming an adult. Immatures are similar in shape to adults, but they are smaller and lack wings. Also called incomplete metamorphosis. Compare to complete metamorphosis.

Solitary parasite. A species in which only one individual develops within a single host.

Specialist natural enemy. A natural enemy that attacks a limited range of usually related host or prey species. It often has behavioral characteristics that allow it to find and attack scarce hosts. Compare to generalist natural enemy.

Species. The group of all genetically and structurally similar individuals that can interbreed to produce fertile offspring and that are different from, and do not normally breed with, individuals from other such groups.

Subspecies. A recognizably distinct subset of a species. Subspecies usually evolve as geographically isolated pockets of a species.

Symbiotic. Organisms of two or more species living together in a dependent relationship.

Thorax. The body region behind the head and in front of the abdomen. Bears the wings and legs.

Trap crop. A pest's preferred host plant adjacent to the crop to be protected. Pests attracted to the trap crop are sprayed or otherwise controlled, preventing them from attacking the main crop.

Vector. An animal that carries a pathogenic microorganism to members of another species. For example, cucumber beetles are vectors of a bacterial wilt of cucurbits and *Culex* mosquitoes are the vectors of West Nile virus.

Photography credits

Photographs were generously loaned by the following contributors. Copyright ownership of all images is retained by the contributors.

Unless otherwise noted, numbers preceding descriptions refer to figure numbers.

Michael F. Antolin, courtesy Michael R. Strand: 63—*Bracon hebetor* larvae

Max E. Badgley: 72—ichneumonid wasp, *Scambus applopapi*; 85—*Trichogramma minutum*; 115—vedalia beetle

Eric H. Baehrecke, courtesy Michael R. Strand: 80—encyrtid wasp, *Copidosoma floridanum,* adults emerging

Joseph Berger, insectimages.org: 62—tachinid fly, *Gymnosoma*; 94—comb-footed spider; 99—two-clawed hunting spider

Ronald F. Billings, Texas Forest Service, insectimages.org: 26—caterpillar hunter, *Calosoma sycophanta*

David Cappaert, Michigan State University, insectimages.org: 35—*Coleomegilla maculata* ; 38—mealybug destroyer larva; 61—tachinid fly, *Arcytas apicifer*; 81—encyrtid wasp, *Encarsia formosa*; 82—parasitized whitefly nymphs; 92—orb-weaving spider; 95—funnel-web spider; 97—jumping spider; 100—crab spider; p.35—Lampyridae beetle; p.35—Lycidae beetle; p.41—Asilidae fly; p.41—Bombyliidae fly

Whitney Cranshaw, Colorado State University, insectimages.org: 37—mealybug destroyer adults; 96—wolf spider; p.35—Meloidae beetle

Tony DiTerlizzi: p.41—Pipunculidae flies

Susan Ellis, insectimages.org: 90—sphecid wasp, *Tachysphex*

Howard Ensign Evans, insectimages.org: 89—sphecid wasp, *Ammophila*

Richard C. Frey, ARBICO: 119—remote-piloted airplane

Ken Gray, Oregon State University: 15—predatory mirid bug; 25—carabid beetle; 39—rove beetle; 43—green lacewing eggs; 44—green lacewings hatching ; 54—hover fly larva; 68—braconid wasp, *Aphidius smithi*; 70—ichneumonid wasp, *Bathyplectes curculionis*; 71—*Bathyplectes* cocoons; 74—*Diadegma* cocoons

Jeffrey P. Gruber, University of Wisconsin: 41—hister beetle, *Carcinops pumilio*

David Hanson, University of Minnesota: p.62—*Binodoxys communis*

Bradley Higbee, Paramount Farming, insectimages.org: 16—predatory mirid nymph; 18—bigeyed bug

Sam Houston: p.41—Pyrgotidae fly

Larry A. Hull, Pennsylvania State University: 36—spider mite destroyers

Jena A. Johnson: 79—encyrtid wasp, *Copidosoma floridanum* adult

Gerald J. Lenhard, insectimages.org: p.35—Elateridae beetle

Richard Leung: p.47—Cynipoidea wasp

Kathleen S. Lieber, University of Wisconsin: 86—*Laelius pedatus* adult; 87—*Laelius pedatus* larvae

Daniel L. Mahr, University of Wisconsin: 10—damselfly; 11—praying mantid; 12—mantid egg case; 14—minute pirate bug nymph; 17—assassin bug; 19—damsel bug adult; 21—spined soldier bug nymph; 22—spined soldier bug adult; 23—spined soldier bug adult; 30—multicolored Asian lady beetle larva; 31—variegated lady beetle pupa; 33—seven-spotted lady beetle ; 34—multicolored Asian lady beetle; 40—hister beetle, *Platysoma cylindrica*; 42—green lacewing adult; 45—green lacewing larva; 46—green lacewing cocoon; 47—green lacewing pupa; 48—brown lacewing adult; 49—brown lacewing egg; 50—brown lacewing larva; 51—predatory midge, *Aphidoletes aphidimyza*; 52—hover fly, *Toxomeris*; 53—hover fly, *Syrphus*; 55—hover fly larva; 56—hover fly pupa; 88—paper wasp, *Polistes fuscatus*; 93—four-jawed spider; 101—harvestman; 105—nematodes in strawberry root weevil larva; 106—nematodes spilling from strawberry root weevil larva; 110—*Bacillus thuringiensis* infection; 111—*Beauveria bassiana* infection; 112—fly killed by *Entomophthora muscae*; 114—frog; 116—hover fly on flower; 120—purchased convergent lady beetles; 121—purchased green lacewing eggs; 122—purchased predatory mites; 123—purchased *Trichogramma minutum*; 124—purchased whitefly parasites, *Encarsia formosa*; 125—purchased parasitized fly pupae; p.35—Cantharidae beetle; p.35—Cleridae beetle; p.41—Dolichopodidae fly

Susan E. Rice Mahr, University of Wisconsin: 29—young lady beetle larvae; 64—*Cotesia glomerata* adult; 65—*Cotesia glomerata* cocoons; 73—ichneumonid wasp, *Diadegma insulare*; 75—chalcidoid wasp, *Pteromalus puparum*; 76—*Pteromalus* larvae; 98—lynx spider; 107—cabbage looper NPV

K.S. Matz: p.47—Chrysididae wasp

B. Newton, University of Kentucky: 27—*Calosoma* larva

Oregon State University: 28—twospotted lady beetle laying eggs; 57—tachinid fly; 58—tachinid eggs; 59—tachinid larva; 60—tachinid pupa; 69—aphid mummy with trapdoor

Russ Ottens, University of Georgia, insectimages.org: 32—convergent lady beetle

J. J. Peterson, USDA: 104—nematode in mosquito larva

Kenneth F. Raffa, University of Wisconsin: 108—gypsy moth NPV; 113—gypsy moth larva killed by *Entomophaga maimaiga*

David B. Rivers and **Matt Meadows**: 77—pteromalid wasp, *Nasonia vitripennis*

M. Scott, Illinois Natural History Survey, insectimages.org: 66—*Macrocentrus grandii*

Merle Shepard, courtesy Michael R. Strand: 67—braconid wasp, *Microplitis croceipes*

Merritt Singleton, University of Wisconsin: 13—minute pirate bug adult; 20—damsel bug nymph; 102—phytoseiid mite, *Amblyseius californicus*; 103—phytoseiid mite, *Phytoseiulus persimilis*

Shawn Steffan, University of Wisconsin: 118—habitat planting next to vegetables

Frederick W. Stehr: 83—eulophid wasp, *Tetrastichus julis*; 84—fairyfly, *Anaphes flavipes*

Louis Tedders, USDA Agricultural Research Service, insectimages.org: 24—twospotted stink bug

Edward Trammel: p.47—Tiphiidae wasp

Ronald M. Weseloh: 78—chalcid wasp, *Brachymeria intermedia*; 91—ant

Paul Whitaker, University of Wisconsin: 117—habitat planting

Index